Caught in the Act – Reflections on Continuing Professional Development of Mathematics Teachers in a Collaborative Partnership

Published by African Sun Media under the SUN PReSS imprint

Copyright © 2019 African Sun Media and the editors

This publication was subjected to an independent double-blind peer evaluation by the publisher.

The editors and the publisher have made every effort to obtain permission for and acknowledge the use of copyrighted material. Refer all enquiries to the publisher.

Views reflected in this publication are not necessarily those of the publisher.

First edition 2019

ISBN 978-1-928480-36-5
ISBN 978-1-928480-37-2 (e-book)
https://doi.org/10.18820/9781928480273

Set in Gothic 72 Lt Bt Light 10/14

Cover design, typesetting and production by African Sun Media

SUN PReSS is an imprint of African Sun Media. Scholarly, professional and reference works are published under this imprint in print and electronic formats.

This publication can be ordered from:
orders@africansunmedia.co.za
Takealot: bit.ly/2monsfl
Google Books: bit.ly/2k1Uilm
africansunmedia.store.it.si *(e-books)*
Amazon Kindle: amzn.to/2ktL.pkL

Visit africansunmedia.co.za for more information.

CAUGHT IN THE ACT

REFLECTIONS ON CONTINUING PROFESSIONAL DEVELOPMENT OF MATHEMATICS TEACHERS IN A COLLABORATIVE PARTNERSHIP

Editors

Cyril Julie
Lorna Holtman
Charles R. Smith

SUN PRESS

CONTENTS

ACKNOWLEDGEMENTS

This book deals with the insights gained in a continuous professional development project dealing with improvement in achievement outcomes in high-stakes mathematics examinations in schools in low socio-economic status environments. The schools and their mathematics teachers are thanked for their sustained voluntarily participation and support. Access to the schools was facilitated through the Western Cape Education Department (WCED) who also willingly supplied us, on requests, with the scripts of the National Senior Certificate: Mathematics examination of the participating schools. The co-operation of the WCED is appreciated. Lastly, this research is supported by the National Research Foundation (NRF) under grant number 77941. Any opinions, findings and conclusions or recommendations expressed in this material are those of the authors and do not necessarily reflect the views of the National Research Foundation of South Africa (NRF SA).

THE EDITORS

DEDICATION

This book is dedicated to the late Jan Persens (1946-2017), Professor in Mathematics and Director of International Relations at the University of the Western Cape, for his tireless efforts for quality teaching of Mathematics at all levels.

CONTRIBUTORS

FAAIZ GIERDIEN works as a mathematics educator in the Department of Curriculum Studies at Stellenbosch University. His research interests are in mathematics education and teacher education, with a particular focus on aspects of continuing professional development and the use of information and communications technology in relation to school mathematics reform.

LORNA HOLTMAN is the Director of the Postgraduate Studies at the University of the Western Cape and an Associate Professor in Science Education. Her research interests are in curriculum studies in Science and research capacity development of post-graduate students for the past 13 years.

CYRIL JULIE is holder of a South African Research Chair Initiative in Mathematics Education at the University of the Western Cape. His research interests are in the development of mathematics teaching, mathematical modelling and political, social, historical and philosophical issues in mathematics education.

MONDE MBEKWA is an Associate Professor in the School of Science and Mathematics Education at the University of the Western Cape. His major research focus is around the use of Information Communication Technology (ICT) in Mathematics Education.

DUNCAN MHAKURE is a Senior Lecturer in the Centre for Higher Education Development at the University of Cape Town (UCT). He has served as a member of the research development group of the Local Evidence-Driven Improvement in Mathematics Teaching And Learning Initiative (LEDIMTALI) project, and his research focus is on mathematics classroom discourses.

ONYUMBE OKITOWAMBA was a doctoral research fellow in LEDIMTALI during the production of this book. His research interest is in assessment of high-stakes mathematics examinations.

MARIUS SIMON is currently a lecturer in Foundation Phase Mathematics in the School of Science and Mathematics Education at the University of the Western Cape. During the production of this book he was a doctoral research fellow in LEDIMTALI. His research interest is in the analysis of learners' ways of working with school mathematics using ethnomethodology and the sociology of science.

CHARLES RAYMOND SMITH is a former senior curriculum planner for Mathematics in the Western Cape Education Department. He later joined LEDIMTALI as a full-time doctoral research fellow. Currently he is a post-doctoral fellow in the project. His research interest is in the continuing professional development of mathematics teachers and the interesection of school mathematics and Mathematics.

JULIANA SMITH is an Associate Professor in the Department of Educational Studies at the University of the Western Cape. Her research interest is action research and school improvement and work-embedded practice of pre-service teachers.

KALVIN WHITTLES is a Senior Lecturer in the Faculty of Education at the Cape Peninsula University of Technology. His research interests include classroom-based research and the study of connections between school mathematics and Mathematics.

ACRONYM LIST

AIMS   African Institute for the Mathematical Sciences

ANA   Annual National Assessments

AP   Acquisitionist perspective

BODMAS   Brackets, Of, Division, Multiplication, Addition, Subtraction

CAPS   Curriculum Assessment and Policy Statement

CHAT   Cultural-historical activity theory

CPD   Continuing Professional Development

CPTD   Continuing Professional Teacher Development

DBE   Department of Basic Education

FET   Further Education and Training

HSIP   High School Intervention Project

ICT   Information Communication Technology

ICTMA   International Community of Teachers of Mathematical Modelling and Applications

IEB   Independent Examination Board

INSET   In Service Education and Training

ISPFTED   Integrated Strategic Planning Framework for Teacher Education

ITM   Intentional Teaching Model

LEDIMTALI   Local Evidence-Driven Improvement in Mathematics Teaching And Learning Initiative

MER   Mathematics education researcher

NAEP   National Assessment of Educational Progress

NBI   National Business Initiative

NRF SA   National Research Foundation of South Africa

NSC   National Senior Certificate

NSES	National School Effectiveness Study
OECD	Organisation for Economic Co-operation and Development
PD	Professional Development
PLC	Professional Learning Communities
PP	Participationist perspective
RECME	Researching Effective CPD in Mathematics Education
RPN	Reverse Polish Notation
SACE	South African Council for Educators
SACMEQ III	Southern and East Africa Consortium for Monitoring Educational Quality
SES	Socio-economic status
TIMSS	Trends In Mathematics and Science Study
UWC	University of the Western Cape
WCED	Western Cape Education Department

INTRODUCTION

Cyril Julie, Lorna Holtman & Raymond Smith

There is a consensus that achievement, as measured by various tests, examinations and surveys, in school mathematics needs improvement. In South Africa, this need is exacerbated by the realisation that achievement in school mathematics is still structured along racial lines. Those at the lowest rung of the apartheid-created racial classification system achieve the lowest; achievement increases as one moves up the ladder of privilege of this classification system. The state, universities and non-governmental organisations instituted various initiatives and projects to address the situation. These initiatives and projects continued with the advent of democracy and a major objective was the improvement of achievement in school mathematics of learners from low socio-economic status environments. The initiatives and projects focused primarily on the enhancement of the school mathematics content knowledge of teachers with some didactical knowledge underpinned mostly by constructivism integrated with the content knowledge offerings.

In 2009, the NRF SA released a concept paper which contained a call for applications for funding. The objectives of the NRF SA Initiative include the following:

1. To improve the quality of teaching of in-service mathematics teachers at secondary schools.
2. To improve the mathematics results (pass rates and quality of passes) in secondary schools as a result of quality teaching and learning.

The accompanying purpose and targets of the FirstRand Foundation South African Maths Education Chairs are to:

1. Develop and implement programmes for the professional development (skills and knowledge) of in-service mathematics teachers at previously disadvantaged secondary schools.

2. Initiate, lead and co-ordinate programmes of research to improve the quality of (a) mathematics teaching and learning, and (b) learner performance at secondary schools (NRF, 2009:3-4).

Furthermore, the document specifically states that the improvement in achievement should be "increases of at least 10% per annum in the number of Grade 12 mathematics passes in each of the schools involved in the programme run by the Chair on average" and "Annual improvements in the quality of the Grade 12 mathematics passes in each of the schools involved in the programme run by the Chair" (NRF, 2009:4).

One proposal that was approved and funded by the NRF was the LEDIMTALI. An underlying feature within LEDIMTALI was that the major objective could only be achieved if there is meaningful collaboration between various role-players in the school mathematics enterprise. These role-players are mathematics educators, mathematicians and curriculum experts based at universities; mathematics teachers; and mathematics curriculum advisors based in the departments of education of the government. LEDIMTALI was consequently a collaborative effort between these role-players, each of whom will bring their experiences and expertise to bear on the project and its development.

LEDIMTALI's major pursuit centred on the development of teaching school mathematics. A further belief was that the development of teaching should have strong ecological relevance. This implies that the reality and contexts of teachers' work in their classrooms should feature strongly in whatever strategies, techniques and tactics are developed to work towards the major objective.

At the outset, LEDIMTALI worked with mathematics teachers in the Further Education and Training (FET) phase (Grades 10 to 12). It soon became clear that these teachers also teach mathematics in the Senior Phase (Grades 8 to 9). Subsequently, the project worked with teachers from both phases in high schools.

This book reports on research over four years of working towards the major objective anchored around the continuing professional development (CPD) of mathematics teachers. It provides insights into how ideas and notions which were developed by the collaborative collective, were or were not appropriated by the participating teachers and the learnings of all participants.

Chapter 1 presents different paradigms for the continuing professional development of teachers. It situates the project within the broader literature of the field and concludes that LEDIMTALI can be conceived as a transformative model of the continuing professional development of school mathematics teaching.

Chapter 2 views the project from a different vantage point. LEDIMTALI is reflected on from a cultural-historical activity theory (CHAT) framework. It focuses primarily on different tools that were at play within the CPD initiative and secondarily on the division of labour between CPD providers and participating high-school mathematics teachers. It illustrates how tools have this uncanny capacity to evolve in work activities. A conclusion reached is that it cannot be accepted that the subjects will unproblematically take on board all tools offered by subjects from one site to use them in another site. This chapter bolsters the ecological relevance of the project.

The mathematical content knowledge of South African mathematics teachers and teachers of mathematics has been described as 'lacking'. This conclusion is based on various tests. These consisted of small-scale studies using opportunistic samples and some large-scale studies using teachers of classrooms which were randomly selected to write national and cross-national systemic tests. **Chapter 3** considers three tests that used large-scale studies. The tests are evaluated against known criteria and other issues linked to the assessment of teachers' content knowledge. It was found that the tests violate some of the criteria for 'good' tests and in some instances the mathematical fidelity of the test items is low. The conclusion is reached that teachers' apparent lack of mathematical content knowledge is more a function of the instruments used rather than of teachers' mathematical content knowledge. It is recommended that proposals flowing from the tests results of teachers' mathematical content knowledge should be viewed with circumspect. It is not denied that teachers have mathematical knowledge gaps, but the chapter argues that these can be identified and addressed during professional development activities.

Chapter 4 deals with the ways that teachers appropriate or not the ideas and notions distributed during CPD activities. It focuses on one particular construct – spiral revision. The chapter concludes that understanding the process as well as the degrees of appropriation has the potential to enhance the effectiveness of teaching development and support.

In **Chapter 5** the focus is on facilitation and mediation by CPD providers. In particular, it emphasises the issue of different positionings of teachers and didactitians. It concludes that mathematics educators involved in CPD initiatives need to constantly search for opportunities to learn what it takes to mediate and to facilitate when they interact with teachers.

Notwithstanding the ecological relevance stance adopted, it was also deemed worthwhile to engage teachers in dealing with issues in the operative curriculum of which they have limited exposure during their education as teachers or in other in-service initiatives. Mathematical modelling was identified as one such issue. The underlying premise is that teachers' primary experience of mathematical modelling is that of contextually embedded word problems. **Chapter 6** provides insight into how practicing teachers dealt with mathematical modelling as content in a short course. It is concluded that continuing professional development initiatives should, at times, also provide practicing teachers with experiences of mathematical work, which does not seemingly have immediate relevance for their practice to widen their perspectives of mathematical work.

Given the nature of continuing professional development of mathematics, there are different operational spaces. Participants engaging in these spaces have different accountabilities and commitments. **Chapter 7** reflects on the space where teachers and professional development providers engage. This is done against the metaphor of a third space where learning is brokered by certain boundary objects and boundary practices.

Chapter 8 continues the theme of the previous chapter. The chapter highlights the role of relational agency in understanding how professional learning is enhanced in collaborative settings of professional learning communities. It reveals how a micro-climate of commonality enables the key affordances for teacher professional learning such as collaborative inquiry, deprivatisation of practice, and dealing with diversity and conflict.

In **Chapter 9** the focus is on learners' ways of working in the high-stakes National Senior Certificate (NSC) Mathematics examination. The actual scripts of examinees are analysed using a quasi-ethnomethodological approach. The chapter argues that such analyses should be done to complement the normal diagnostic report provided by the Department of Basic Education on the NSC examination. This suggestion is accompanied with a caution that teachers simply do not have the time to do the kind of analysis presented and suggests that CPD providers should do this analysis using the actual responses of learners' scripts produced from high-stakes examination context.

Chapter 10 addresses the primary objective of improvement in achievement and the quality of passes in the NCS Mathematics examinations. It addresses how examination-driven teaching eventually underpinned the LEDIMTALI and the ostensible impact of this approach on achievement and its quality. Various models are used to analyse four years of examination results of the participating schools. The analyses demonstrate that, on average, there was a definite progression in a positive direction. This progression is, however, not pronounced. This is explained using various postulates which offer reasons for the gradualness of transformation.

To conclude the *Caught in the act* in the title captures the ecological relevance stance of LEDIMTALI. The 'catching' is about the identification of the positive aspects of teachers' work in their classrooms, the appropriation of the 'goods' distributed during CPD sessions, the investigation of the actual work of learners, the learnings of CPD providers and research related to the objective of the project. Everything that is caught is translated for use by all participants to push towards improved learner performance. This, in our opinion, is to counter a 'pedagogy of poverty' (Haberman, 1991) which, it can be argued, dominates many CPD initiatives in South Africa. This is also sometimes the case for initiatives which claim to not adhere to a deficit model of CPD by pronouncing that the initiatives work with teachers from where they are and not working on teachers. Our overall contention is that to enhance the achievement in high-stakes mathematics we must, as a first approximation, work with what teachers bring to an initiative and productively develop the positives that are caught in the acts.

References

Haberman, M. 1991. The pedagogy of poverty versus good teaching. *Phi Delta Kappan*, 73:290-294.

NRF (National Research Foundation). 2009. Concept paper: *FirstRand Foundation South African Maths Education Chairs Initiative*. Pretoria: NRF.

|01|

INITIAL INSIGHTS INTO A CONTINUING PROFESSIONAL DEVELOPMENT PROJECT FOR MATHEMATICS TEACHERS

Monde Mbekwa, Kalvin Whittles, Lorna Holtman & Juliana Smith

Introduction and background

This chapter reports on a CPD project, LEDIMTALI based at a higher education institution in the Western Cape Province of South Africa. It provides insights into the initial stages of how the project team conceptualised CPD given the inputs from various stakeholders and models found in the literature.

The partnership is aimed at increasing the number of learners taking mathematics as an examination subject in the FET phase of schooling in South Africa. The partnership also focuses on improving the levels of performance of learners in the NSC Mathematics examination. In particular, as stated in the overall concept document by the NRF SA, one of the objectives is "To improve the mathematics results (pass rates and quality of passes) in secondary schools as a result of quality teaching and learning" (NRF, 2009:3). This objective makes it clear that attention should be given to improve the performance in mathematics of learners from low socio-economic status environments. In essence, this is part of the broader social transformation project of school education in South Africa where achievement in high-stakes national examinations still mirrors the pattern prevalent during the colonial and apartheid era.

One of the interventions by the national government was the Dinaledi Project. This initiative was launched by the former Department of Education in 2001 with 102 schools under a National Strategy for Mathematics, Science and Technology Education (National Business Initiative (NBI) 2009). The main purposes of this strategy were to increase the number of learners taking Mathematics and Science as examination subjects and to improve learner performance in those subjects. In 2015, more than 550 schools were participating in the Dinaledi Project. According to a departmental spokesperson, the Dinaledi Project not only focuses on learners' performance in mathematics but also the augmentation of teacher mathematical content knowledge (Buthelezi, 2012). This focus on teachers' content knowledge is a widely held assumption that teacher subject knowledge is an important factor in learner performance.

As part of this initiative, the Department of Basic Education (DBE) has initiated an Adopt-A-School subproject which is a private sector support programme for some Dinaledi schools. Standard Bank has adopted more than 100 Dinaledi schools and supports them financially with a grant of R50 000 each. The grant was to be used to appoint a subject specialist to help with the teaching of mathematics and science. It is instructive to note that despite this initiative and focused, direct intervention by the DBE in schools to improve performance in mathematics, the results of learners' performance in mathematics did not show much of an improvement (NBI, 2009).

Departmental continuing professional development initiatives

The DBE recognises the importance of continuing professional development of teachers to enhance the quality of provision of education in South Africa. In pursuance of this objective, the South African Council for Educators (SACE) was tasked with the responsibility of registering and administering general CPD programmes for teachers. A pilot study was subsequently initiated and was conducted in 146 schools in all nine provinces (SACE, 2007, 2012). Currently, SACE are in the process of implementing the CPD system across all schools in the country. To strengthen its resolve to attend to the CPD needs of all teachers, the DBE formulated the Integrated Strategic Planning Framework for Teacher Education (ISPFTED) in South Africa 2011-2025. The ISPFTED envisages a system of CPD where teachers take responsibility for their professional development. According to this planning framework, teachers may do this by:

Ø learning how to identify gaps in knowledge through (a) interpreting learners' results in national and other assessments, and (b) taking user-friendly online and/or paper-based diagnostic tests in specific subject/learning areas;

Ø actively learning with colleagues in professional learning communities (PLCs);

Ø accessing funding to do quality-assured courses that are content-rich and pedagogically strong and that address their individual needs;

Ø understanding the curriculum and learning support materials, preparing lessons and delivering them competently; and

Ø signing up with the SACE Continuing Teacher Professional Development management system and achieving the targeted number of Professional Development (PD) points.

The WCED, on the other hand, has introduced several initiatives to improve matric performance, inter alia, a 'winter school' for Grade 12 learners and a High School Intervention Project (HSIP) targeting underperforming schools in the province (WCED, 2012). HSIP responds to the analyses of the matriculation results of schools that obtain less than a 60% pass rate in the final examinations.

Besides the formal initiatives around CPD, universities and other providers are also striving for these goals, and LEDIMTALI is one such project.

Continuing professional development for teachers

This section conceptualises continuing professional development for teachers and discusses the features of effective CPD for teachers

CPD defined

There are many ways in which CPD may be conceptualised. For this book, the following conceptualisation of CPD is adopted: The CPD of teachers refers to those activities aimed at improving the knowledge and skills of teachers through orientation, training and support (Coetzer, 2001; Lessing & De Witt, 2007).

This chapter is also informed by the two descriptions of CPD. Day proposes the following:

> Professional development consists of all-natural learning experiences and those conscious and planned activities which are intended to be of direct or indirect benefit to the individual, group or school and which contribute through these to the quality of education in the classroom. It is the process by which, alone or with others, teachers review, renew and extend their commitment as change agents to the moral purposes of teaching, and by which they acquire and develop critically the knowledge, skills and emotional intelligence essential to good professional thinking, planning and practice with children, young people and colleagues through each phase of their teaching lives. (Day 1999, cited in Orgoványi-Gajdos, 2016:72-73)

Bolam describes CPD as activities teachers engage in after their initial teacher training as:

> Those education and training activities engaged in by secondary and primary school teachers and principals, following their initial professional certification and intended mainly or exclusively to improve their professional knowledge, skills, and attitudes in order that they can educate children more effectively. (Bolam, 1982:14)

There is a common thread that runs through these descriptions:

Ø Teachers engage in particular kinds of activities which may be conceived of as professional learning opportunities.

Ø These activities are both formally or informally orchestrated and may be planned or incidental learning opportunities.

Ø These activities encompass a multiplicity of modalities.

Ø The aim is to improve or change knowledge and skills as well as dispositions or beliefs.

Ø The motivation is better learning outcomes by enriching the pedagogical practices of teachers.

CPD is an important mechanism for enhancing the quality of teaching and learning in mathematics classrooms. CPD may include activities, such as attending conferences, workshops, formal subject meetings and discussions which do not lead to a formal qualification. However, as part of CPD, teachers can follow credit-bearing courses which can culminate in a qualification. This is in agreement with Back, Hirst, De Geest, Joubert and Sutherland (2009:11), who view the provision of CPD to include a range of opportunities, such as "whole school training days, team planning opportunities, joint teaching, peer observation, work shadowing, residential working groups, and local and national conferences and networks".

Features of effective CPD for teachers

Research has shown that certain forms of professional development in education have little to no noticeable impact in classrooms (Darling-Hammond & McLaughlin, 2011; Nolan & Hoover, 2004; Peery, 2004). Researchers, for example, Ball & Cohen (1999); Borko (2004) and McLaughlin & Talbert (2001), believe that high-quality CPD for teachers meet the following criteria:

Ø It grounds teachers in both content and pedagogy.

Ø It allows teachers to practice new ideas in contexts similar to their classrooms.

Ø It is sustained over time.

ø It offers a community of peers and coaches that provide support as well as opportunities to collaborate.

ø It is resource-rich.

Noting the consensus in the literature about what renders CPD for teachers effective, the LEDIMTALI project is predicated on the following core beliefs about CPD.

TABLE 1.1   Core beliefs for CPD featured in the LEDIMTALI project

Core beliefs for CPD	LEDIMTALI project
CPD is critical for improved student learning through the development of classroom teaching practices.	Collaborative reflecting on and sharing of teaching practices. Collaborative planning, discussing, unpacking, interpreting and inquiring about the mathematics curriculum. Discussing and sharing of assessment practices. Learning about appropriate learning theories and pedagogies relevant to mathematics teaching and learning. Dealing with diversity and learners' attitudes towards mathematics.
CPD is most effective when it is collaborative and collegial.	Developing respectful ways of working amongst and between the stakeholders (mathematics teacher educators, mathematicians, mathematics teachers and mathematics curriculum advisors) for the enhancement of teaching mathematics. An ethos of trust and integrity is central to strengthen the collaboration.
Collaborative work should involve inquiry and problem-solving in authentic contexts of daily teaching practices.	Project workers visit and observe the operationalisation of the teachers' implementation or not of the ideas and notions distributed in workshops and institutes. Furthermore, teachers report on their practices at workshops and institutes as a form of sharing and reflection.

Models of CPD

Kennedy (2005) proposes a framework which identifies nine key models of CPD and explores the forms of knowledge that can be developed through any particular model. The power dynamics inherent in the individual models are examined and the extent to which CPD is conceptualised and promoted, either as an individual attempt or as a collaborative attempt to support transformative practice.

The proposed models are as follows: training, award-bearing, deficit, cascade, standards-based, coaching/mentoring, community of practice, action research and transformative. These nine models are not mutually exclusive but are rather are an endeavour at identifying key characteristics of different types of CPD. This is aimed at a deeper understanding of the models and promoting dialogue about their purpose.

These CPD models can be collapsed into three broad categories – transmission, transitional, and transformative. The transmission model exemplifies the transfer of knowledge without allowing the teacher to make any contribution to the discussion in terms of his or her experience. The transitional model allows for more flexibility in terms of which teachers' professional capacity is built by sharing their experiences with other teachers and then trying to make sense of these experiences. The transformative model allows for reflection of practice in order to change practice through action. Table 1.2 illustrates these models.

TABLE 1.2   Spectrum of CPD Models

Model of CPD	Purpose of model	
ø Training model ø Award-bearing model ø Deficit-model ø Cascade model	Transmission	Increasing capacity for professional autonomy
ø Standards-based model ø Coaching/mentoring model ø A community of practice model	Transitional	
ø Action research model ø Transformative model	Transformative	

Source *Adapted from Kennedy (2005:248)*

Postholm suggests that professional learning for teachers may take place in various ways. She suggests that teachers could attend and participate in workshops as well as doing long and/or short university courses. Also, the learning could be job-embedded by teachers continually reflecting on their practice and the learning of their students. Postholm also suggests that professional learning will take place by observing colleagues in practice and giving interactively giving feedback. Finally, she proposes that the learning will take place in informal conversations with colleagues (Postholm, 2012:406).

Research approach

The research reported on is situated in a qualitative interpretive framework. In this framework, the quest is to provide descriptions of the phenomena under scrutiny. At the outset, it was accepted that these descriptions should be credible, transferable and dependable as is the case for all kinds of interpretive research. The credibility of the collected data was striven for through member checking, including the teachers whose lessons were observed. Another source of data was information relayed by participants during informal conversations in workshops and institutes. Such information was shared with other members of the research team and double-checked with the conveyors of the information. In addition, the end-of-year questionnaire contained an item referring to 'new' knowledge participants gained. This ensured the credibility of such data through triangulation.

The transferability was ensured by reporting studies in sufficient detail so that others could, where appropriate, relive the experiences the research team went through in reporting "so candidly that it justifies itself, and that this experience can be transmitted to others to become like their own experience" (Freudenthal, 1991:161). It is hoped that this chapter conforms to this kind of reporting.

Lastly, dependability was achieved by archiving all matters pertaining to the project by the primary project coordinator and making these electronically available to all members of the research team. This ensures that the primary data are available to any interested parties to re-analyse and retrace the processes followed by the research team.

Multiple sources of data were collected. These included observation notes and audio-recordings of classroom lessons, after-school workshops and institutes, video and audio-recordings of institutes, incidental and informal conversations on issues of importance to projects and two questionnaire administrations. One questionnaire was administered at the end of the first institute and the other at the final workshop for the year. The first-mentioned questionnaire requested teachers to respond to the three questions:

1. What did you not find useful or beneficial in terms of your expectations?
2. What did you find useful or beneficial in terms of your expectations?
3. What changes would you like to see in future workshops?

The year-end questionnaire was an adaptation of the online questionnaire used by the Researching Effective CPD in Mathematics Education (RECME) (Joubert, Back, De Geest, Hirst & Sutherland, 2009). Data analysis was done against the framework articulated above and competitive argumentation of research team members regarding the placement of data in the particular categories of the framework. Note that at all times ethical considerations were always honoured.

CPD Model of the project under discussion

The project under discussion, although institutionally driven, is conceptualised differently. This is a project where mathematics educators, mathematicians, mathematics teachers and mathematics curriculum advisors work collectively and collaboratively to develop and achieve quality teaching of mathematics. The initiative is premised on the belief that such collective and collaborative work can lead to learners achieving at their highest potential in mathematics. The interaction of the major stakeholders provides a space in which trust is developed towards a common purpose.

The CPD model in LEDIMTALI may be conceived as a transformative model. The model did not start with pre-conceived gaps or deficits in teachers' practices. Instead, it proceeded from the supposition of helping teachers to do what they are doing better. Collaboratively, the notion of solution-seeking to problems emanating from real classrooms lead to a model of CPD which may be characterised by the representation of teacher change developed by Clarke and Hollingsworth.

In the representation shown at Figure 1.1, the solid arrows represent the process of enaction; the broken arrows represent the process of reflection. This implies that the way that one domain exerts influence on another is through these two processes. This was indeed our experience of the LEDIMTALI project, namely what emanated from our teacher institutes was taken to the classroom by the teacher and the teacher reported back on the implementation during the next institute.

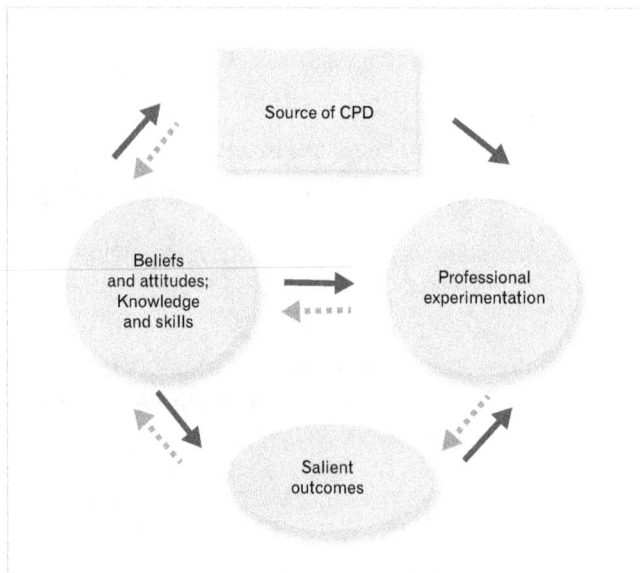

FIGURE 1.1   Model of CPD

SOURCE   CLARK & HOLLINGSWORTH (2002:951)

The domain described as the source of Continuing Professional Teacher Development (CPTD) may be characterised as the external domain and is influenced through a circular pathway through running cyclically through the other domains. This implies that the source of CPTD was never pre-configured but responded to the experiences of participants.

The professional experimentation oval represents the domain of practices. The salient outcomes oval represents the domain of consequences. The beliefs, attitudes, knowledge and skills represent the personal domain.

LEDIMTALI as a professional learning community

The CPTD model intentionally adopted the operational conventions of a PLC. According to Hunter and Black (2011:94), PLCs hold much promise in "developing sustainable networks of teachers of mathematics who engages in developing effective pedagogy". So, as the schools and teachers became engaged in building collaborative work cultures, the term 'learning organisations' was replaced by professional learning communities (Dufour & Eaker, 1998:15). The advantages of following the prescripts of a PLC as the operational tradition are fivefold.

1. Shared practices, which are the result of a collaboration that leads to shared routines, artefacts, conventions and histories.
2. Shared meanings, which are constructed during collaborative discourses and produce a language by which participants can talk about their changing experiences and abilities, and to explain the way things are in their classrooms and beyond.
3. Shared competence, which occurs when participants in a PLC develops practice, that is, personal and shared competences in order to do what needs to be done.
4. A sense of community which is created by the way that relational agency is enacted, and the way participants talk about the social arrangements they employ.
5. A sense of identity, which results from being immersed in a shared history and notions of who they are and how they fit into the PLC.

Initial insights: Observations and discussion

The LEDIMTALI CPD model embraces the five core tenets of a professional learning community, as encapsulated in Table 1.3.

TABLE 1.3   Core tenets of a PLC as enacted in LEDIMTALI

Core tenets of a PLC	Enactment in LEDIMTALI
Shared vision, norms and values	Improving participation and performance in mathematics by implementing intentional teaching and aligned strategies.
Supportive leadership and shared leadership	Facilitation by various participants including teachers themselves.
Reflective dialogue	Teachers sharing practices and reflecting on lesson observations.
Collaborative inquiry	Participants engaging in collective solution-seeking. Teachers experimenting with suggested solutions and giving feedback during Institutes
Deprivatised practice	Sharing non-evaluative feedback on observed lessons and/or videotaped lessons.

Source **Servage (2008:63)**

From the data collected, we arrived at the findings, which can be organised into categories adapted from a typology of outcomes formulated by Harland and Kinder (1997). The subsections that follow discuss these insights into the outcomes of the project.

Material and resource provisioning outcomes

In this project, resource provisioning has taken the form of a book by Watson and Mason (1998) given to all the participating teachers in the project as well as a laptop computer given to each school. Workshop material in the form of selected mathematics questions and presentations has also been given to the participants. This included exemplar question papers for use by teachers in their schools. The provision of these materials is deemed to be a motivating factor for the continued participation of teachers in the project.

Informational outcomes

Informational outcomes refer to the provision of relevant information to the participants regarding the curriculum and processes involved in curriculum development and assessment. In the project under discussion, a decision was taken to set a common mathematics examination paper for the Grade 10 end-of-year examination for all participating schools. Many of the participating teachers were not exposed to setting an examination question paper from the perspective of an external examination. It was, therefore, insightful to be involved in this process and to note the technical details regarding how a paper is set including the distribution of cognitive levels within the question paper as per the recommendation of the Education Department.

While discussing the setting of the question paper, the teachers indicated that had they known beforehand that a common question paper would be set; they would have prepared themselves better. For instance, teachers feel that had they known that the project would propose a common paper, issues such as "curriculum stripping and curriculum exclusion" could have been attended to earlier in the year. This implies that they would have focused on those topics specifically mentioned for examinations in the curriculum documents and excluded other topics deemed to be of lesser importance. This points to the examination as both an assessment tool and a curriculum 'organiser'. Another example of curriculum information negotiation in the project occurred during the planning for the Grade 11 mathematics workshops in 2013. Issues under discussion were the integration of the mathematics curriculum and the vertical linkages of parts from Grades 10-12, which would inform the discussion in the project in 2013. The following are initial insights we have observed in the project in 2012.

New awareness

New awareness refers to the adoption of a new paradigm on previously held knowledge and beliefs by participants on the curriculum or parts of the curriculum. Harland and Kinder (1997:74) refer to this outcome as new awareness and describe it as "a perceptual or conceptual shift from previous assumptions of which constitutes (sic) the appropriate content and delivery of a particular curriculum area."

Value congruence and impact on practice

Value congruence and impact on practice is defined by Harland and Kinder as "the personalised versions of curriculum and classroom management which inform a practitioner's teaching and how these … come to coincide with the In Service Education and Training (INSET) providers" messages about 'good practice' (Harland & Kinder, 1997:73). In this chapter, value congruence, for instance, can be illustrated by the agreement between the teachers, INSET providers and the curriculum advisers that homework is necessary to consolidate previously done work as an assessment tool and as an introduction to new work. Teachers, nevertheless, did not have a consensus on the frequency of giving of homework. Some had a tradition of giving homework daily and others gave it twice or thrice weekly. The INSET providers suggested that instead of homework, a strategy to replace homework should be adopted. This is the notion of spiral revision, which means that at the beginning of every lesson a short exercise of about five to ten minutes is given to learners to revise previously done work. This short exercise is marked before the new lesson can proceed. Another issue that can be regarded as value congruence is that of summative assessments of mathematics classes at the end of the year. All participants agreed on this. A change was proposed on the issue of a final assessment. There was consensus among participants to have all Grade 10 learners in the participating schools write a common examination. This examination was written and marked by participating teachers.

Affective outcomes

Affective outcomes refer to those activities of the project that generally elicited positive responses and motivation of participants. These activities include examination setting, which was a collective exercise of all participants, drawing up memoranda, and a workshop on the assessment given by one of the teachers. The latter allowed teachers to see themselves through the presentation given by 'one of them' as taking charge of their learning. They also had a sense that they were more informed about the curricular process and hence more receptive of the CPD project.

Motivational and attitudinal outcomes

The project has created a forum in which issues pertinent to teaching and learning are discussed. The atmosphere created is one of consensus, collaboration and shared meaning. This was enhanced by short motivational sessions where participants shared positive classroom experiences from their teaching of school mathematics. The relationship between teachers, the teacher educators, mathematicians and the curriculum advisers is one of mutual respect and mutual support. The teachers do not use the interaction with curriculum advisers and teacher educators as sessions for complaints about their schools and the education department but to discuss issues on the agenda of the project.

Knowledge and skills

One of the most outstanding features of the project is a sense that teachers are getting more knowledge and skills from the input of mathematicians and INSET providers in the project. Mathematicians and mathematics teacher educators in the project elaborated on some pure mathematics concepts and teachers felt that they had gained new knowledge. For instance, a few teachers approached the project chairperson to confess that they were not aware that one of the axioms for the congruency of a triangle, namely two sides and an included angle implied that the angle had to be between the equal sides. This was new knowledge to them. In one of the workshops, a presentation on elementary number theory allowed teachers to gain new knowledge on issues of divisibility.

What has been noted in the project is that during general discussion only certain voices are heard, but in group discussion, the voices of the younger teachers emerge, and they assert themselves. The teachers are not afraid of making mistakes but are willing to challenge the more experienced mature members of the project.

Institutional outcomes

The regular meetings, workshops and teacher institutes have created an opportunity and space for the teachers to plan for improvement of curriculum delivery and integration, including sequencing of the curriculum in Grade 11 in the following year. This also includes setting the ground for continuing and summative assessments of the Grade 11 curriculum.

Conclusion

The foregoing discussion highlights some of the emerging issues as the CPD project for mathematics teachers in the FET phase evolves. In 2012, the project focused on Grade 10, however, it currently includes Grades 11 and 12. This is work in progress and gives some initial observations of what has transpired. A more complete picture will emerge as the project evolves to its conclusion. What is beginning to emerge in this project is that there is a discernible development of cohesion amongst the participating stakeholders and growing confidence and trust as the four groups of mathematics practitioners meet and interact.

References

Back, J.; Hirst, C.; De Geest, E.; Joubert, M. & Sutherland, R. 2009. *Final report: Researching effective CPD in mathematics education (RECME)*. Sheffield, UK: National Centre for Excellence in Teaching of Mathematics (NCETM).

Ball, D.L. & Cohen, D.K. 1999. Developing practice, developing practitioners: Toward a practice-based theory of professional education. In: G. Sykes & Darling-Hammond (eds.), *Teaching as the learning profession: Handbook of policy and practice*. San Francisco: Jossey Bass. pp.3-33.

Bolam, R. 1982. INSET for Professional Development and School Improvement, *British Journal of In-Service Education*, 9(1):9-21. https://doi.org/10.1080/0305763820090103

Borko, H. 2004. Professional development and teacher learning: Mapping the terrain. *Educational Researcher*, 33(8):3-15. https://doi.org/10.3102/0013189X033008003

Buthelezi, N. 2012. *Inside efforts to turn the tide on SA's poor maths and science performance*. Available: http://bit.ly/2Zm7eGg (Accessed 5 December 2012).

Clarke, D. & Hollingworth, H. 2002. Elaborating a model of teacher professional growth. *Teacher and Teaching Education*, 18:947-967. https://doi.org/10.1016/S0742-051X(02)00053-7

Coetzer, I.A. 2001. A survey and appraisal of outcomes-based education (OBE) in South Africa with reference to progressive education in America. *Educare*, 30(1):73-93.

Coffield, F. 2000. *The necessity of informal learning*. Bristol: Policy Press.

Darling-Hammond, L. & McLaughlin, M.W. 2011. Policies that support professional development in an era of reform. *Phi Delta Kappan*, 92(6):81-92. https://doi.org/10.1177/003172171109200622

DuFour, R. & Eaker, R. 1998. *Professional learning communities at work: Best practices for enhancing student achievement*. Bloomington, IN: Solution Tree.

Freudenthal, H. 1991. *Revisiting mathematics education*. Dordrecht: Kluwer Academic Publishers.

Harland, J. & Kinder, K. 1997. Teachers' continuing professional development: framing a model of outcomes. *Journal of In-Service Education*, 23(1):71-84. https://doi.org/10.1080/13674589700200005

Hunter, J. & Back, J. 2011. Facilitating sustainable professional development through lesson study. *Mathematics Teacher Education and Development*, 13(1):94-114.

Joubert, M.; Back, J.; De Geest, E.; Hirst, C. & Sutherland, R. 2009. Professional development for teachers of mathematics: Opportunities and change. In: V. Durand-Guerrier, S. Soury-Lavergne & F. Arzarello (eds.), *Proceedings of CERME,* 6:1761-1770. Lyon, France: Institut National de Recherche Pédagogique. Available: http://bit.ly/2Ldxnx6 (Accessed 11 June 2014).

Kennedy, A. 2005. Models of continuing professional development: A framework of analysis. *Journal of In-service Education*, 31(2):235-250. https://doi.org/10.1080/13674580500200277

Lessing, A. & De Witt, M. 2007. The value of continuous professional development: teachers' perceptions. *South African Journal of Education*, 27(1):53-67.

McLaughlin, M.W. & Talbert, J.E. 2001. *Professional communities and the work of high school teaching*. Chicago: Chicago University Press.

NBI (National Business Initiative). 2009. *The Dinaledi Schools Project Report June 2009*. Available: http://bit.ly/2NH26Ft (Accessed 29 November 2012).

NRF (National Research Foundation). 2009. *Concept paper: FirstRand Foundation South African Maths Education Chairs Initiative*. Pretoria: NRF.

Orgoványi-Gajdos, J. 2016. *Teacher's professional development on problem solving: Theory and practice for teachers and educators*. Rotterdam/Boston/Taipei: Sense Publishers. https://doi.org/10.1007/978-94-6300-711-5

Postholm, M.B. 2012. Teachers' Professional Development: A Theoretical Review. *Educational Research*, 54:405-429. https://doi.org/10.1080/00131881.2012.734725

SACE (South African Council for Educators). 2007. *Continuing professional development System*. Available: http://bit.ly/2ZkyV22 (Accessed 4 December 2012).

SACE (South African Council for Educators). 2012. *The status of the CPTD management system: a report based on the pilot.* Available: http://bit.ly/2NFMzWu (Accessed 20 September 2013).

Servage, L. 2008. Critical and transformative practices in professional learning communities. *Teacher Education Quarterly*, Winter:63-77.

Watson, A. & Mason, J. 1998. *Questions and prompts for mathematical thinking.* Derby, UK: Association of Teachers of Mathematics.

WCED (Western Cape Education Department). 2012. *Educational programmes aimed at improving matric performance.* Media statement, 11 July. Available: http://bit.ly/30TbLfK (Accessed 3 April 2015).

|02|

A CONTINUING PROFESSIONAL DEVELOPMENT PROJECT FOR MATHEMATICS TEACHERS: AN ACTIVITY-THEORY PERSPECTIVE

Monde Mbekwa & Cyril Julie

Introduction

CPD for teachers shares the characteristic with initial teacher education that teacher educators from one site (normally a teacher education institution, department or faculty of education) working with teachers from another site, normally schools. Most of the time this enhances aspects of teaching. These different sites have their own commitments and accountabilities. Hence, there are tensions and contradictions as well as affordances and discontinuities within CPD for teachers. For example, not everything that is offered by teacher educators is accepted by teachers. It is also not the case that teacher educators would address issues which teachers feel are their immediate needs.

The system of CPD for teachers is a complex one and needs an appropriate set of descriptive mechanisms to capture its complexity, albeit that it can never be done exhaustively. The mechanisms must be robust enough to capture the range of socio-cultural relations operative during CPD activities. Activity theory, and its later development into CHAT, is widely deemed to offer a useful framework to describe teacher education initiatives.

Wilson (2014:21) describes some studies to demonstrate "how CHAT has been used in different research studies to analyse aspects of teacher education." She does not use other empirical data, however, as is inherent in the quotation above, she refers to the research done by others using CHAT. The studies Wilson cites focus on initial teacher education, and she concludes that the use of CHAT has much to offer to facilitate change.

McNicholl & Blake (2013) use CHAT to consider the tensions and contradictions teacher educators experience in higher education institutions. The tensions and contradictions are primarily found between the need to do research and enhance compliance to educate practice-ready teachers. They couch their work in the broader political issue in England and Scotland where there are offered imperatives to shift the education of prospective teachers from universities to schools. McNicholl & Blake conclude their study by stating:

> Although teacher educators are well aware of the need to do research, and would generally want to do research, there tends to be little acknowledgement by universities of their different background and specific developmental needs … The preponderance of artefacts from the professional context indicates a highly conscientious commitment to teaching at the expense it seems of time for research. Perhaps this level of commitment is more than is necessary within partnership teacher education, and reveals one aspect of the difficulty of adapting to a different culture, or of even making sense of the crucial difference that research might make to the balance of work which must increasingly meet multiple tests of rigour and relevance in intersecting research, practice, public scrutiny and usefulness. (McNicholl & Blake, 2013:22-23)

An earlier study within the CHAT framework is that by Roth & Tobin (2002). Their study dealt with the capacity of prospective teachers to teach in schools characterised by diverse social, economic and cultural identities which the pre-service teachers are used to and are comfortable with. They argue that the difficulties teachers experience are an outcome of the shortcomings of their "education courses, including … science teaching methods course" (Roth & Tobin, 2002:118), which lead them to redesign their education courses by foregrounding co-teaching. A major conclusion they reach is that "Co-teaching establishes new relations between the stakeholders from both sides of the school-university partnership, in class and in meetings designed to allow for learning through reflection on action." (Roth & Tobin, 2002:133).

Different to the research studies referred to above, Wake, Swan and Foster (2015) report on a study regarding practicing teachers' involvement in lesson study. They pursued the question "How do we support and facilitate teachers in their dual roles of being teachers and learners?" (Wake et al., 2015:244).

The researchers and the teachers planned a problem-solving lesson. Teachers further developed the initial planning to be more classroom-implementable according to their assessment of what features must be altered to develop a draft lesson. The researchers advised on particular issues of the lesson plan. One of these issues focused on was the anticipated solutions learners might give. Teachers found this difficult and they addressed it by incorporating the problem that would be dealt with in advance of the actual lesson research lesson. This was a 20-minute lesson in which learners had to solve a problem unaided. The responses provided by the learners were analysed to obtain a set of anticipated responses and the questions and prompts to support the learning that would be focused on in the actual lesson.

Another issue related to the recognition of learners' progress. Using their expertise in lesson design, the researchers, co-developed with the teachers a progression scheme of learner achievement. Eighteen people observed the implementation. These included teachers, academics (from the UK and Japan) and PhD students whom all participated in the post-lesson discussion as is the practice in the lesson-study approach. One of the insights gained from situating the study in the CHAT perspective was the role the meditation tools, particularly the anticipated responses and the progression scheme, played in lesson preparation and post-lesson discussion. The authors conclude that their "analysis suggests that artefacts, such as the anticipated issues and progression tables, are important in facilitating this collaboration and joint responsibility as well as stimulating professional learning." (Wake et al., 2015:251). The professional learning is not only those of teachers but also of the researchers.

As is the case with the aforementioned study, this chapter deals with work with practicing teachers. From a CHAT perspective, it focuses on the meditational tools, their emergences, their use and non-use in a project, and the LEDIMTALI. It also describes how the division of labour manifested itself in the project. The next section provides a brief description of CHAT.

Cultural-historical activity theory and activity theory

This section explores the concepts of CHAT and activity theory and discusses activity systems as the essence of activity theory. The section explores expansive learning and the use of CHAT as a tool of analysis.

Activity systems as the essence of activity theory

Cultural-historical activity theory, the analytical framework adopted for this chapter, is a reformulation of the Vygotskian socio-cultural-historical theory by

Engeström (1987), which conceptualises human action as being embedded in artefact mediated activity systems. The type of system considered here is that of continuing professional development of a mathematics teaching project viewed as an activity whereby people are jointly engaged in a process of teaching development. The unit of analysis of all these systems is activity. Simon Goodchild defines 'activity' as "The link between the acting person and the object on which she acts to achieve some desired outcome" (Goodchild, 1997:27). Hence, activity systems are seen as social practices of people involved in some action to improve the practice of all involved in the situation. As Aleksei Leont'ev suggests:

> [Activity theory] …is concerned with the activity of concrete individuals, which takes place in a collective – i.e. jointly with other people – or in a situation in which the subject deals directly with the surrounding world of objects…the human individual's activity is a system of social relations. It does not exist without these relations. (Leont'ev, 1981:47)

One can conclude from the above formulation that a critical dimension of activity systems is their social nature and most importantly the notion of mediation through the use of tools and signs. Tools reside in the physical realm, whilst signs reside in the psychological domain (Vygosky, 1978). Examples of tools in an educational activity are concrete objects, such as pens, books and computers. Signs, on the other hand, can be exemplified by human language and thought.

In terms of Vygotsky's model of activity, the subjects are the persons who act, the object is the intended goal of the activity, and the mediating artefact is the moderating device, which determines how the action is performed. Wake, et al., (2015) exemplifies the notion of an object and refers to the mathematics classroom as an activity system and the learning of mathematics as the object of activity. Vygotsky's model was modified by Engeström (1987) to include other components such as rules, the community and division of labour, which are in a dialectical relationship with one another. Figure 2.1 shows Engeström's triangular model for an activity system.

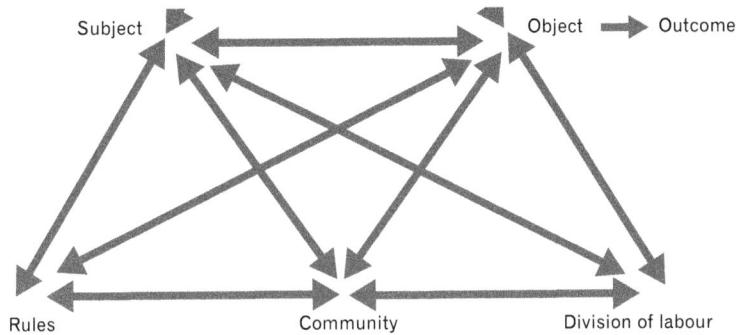

FIGURE 2.1   Triangular model of an activity system
Source   Engeström (1987:78)

The unit of analysis in Engeström's theory is activity as previously defined by Goodchild. The elements of the activity system work collaboratively as agents of change and constitute interactive networks.

Similarly to Vygotsky (1978), Engeström (1987) defines the subject in the activity system as the people who are involved in an activity, the object as the goal or motive force of the activity, and the mediating artefacts or tools as the psychological or physical means used to assist the subject to achieve outcomes. The other components that complete the activity system are the community, the people who interact in the activity system and who share the same object of the activity, and rules, which refers to modes of behaviour or ways of how to work within the system. Another element is the division of labour, which refers to the distribution of roles for members of the system. Engeström avers that "as more actors join in a collaborative analysis and modelling the zones of proximal development are initiated and carried out" (Engeström, 2010:78). The kind of learning generated by these networks is known as expansive learning, which is briefly discussed in the next section.

Expansive learning

Engeström, (1987, 1993, 1999, 2010) postulates that expansive learning is a way of resolving the tensions inherent in activity systems. He defines it as a way of determining the essence of an object "by tracing and reproducing theoretically the logic of its development, of its historical formation through the emergence and resolution of its inner contradictions" (Engeström, 1999:382).

In simple terms, expansive learning develops when people begin to analyse their learning, including the intervention of others in the activity system intending to develop and connect with other activity systems. It arises out of the need to expand the learning of the participants of an activity system to generate new learning and innovation. Expansive learning begins from a Davydovian understanding of learning as developing from the abstract to the concrete through a system of learning cycles. It manifests itself as a transformation of the objects of activity as subjects begin to question and analyse the activity system seeking out explanatory models of new learning, implementation and then a reflection of the outcomes and their consolidation.

According to Engeström, expansive learning should be seen as a complex process of resolving tensions within activity systems. It is also a way of creating new knowledge by questioning through what he refers to as "charting the zone of proximal development that needs to be traversed in order to move beyond the existing contradictions" (Engeström, 2010:81).

Cultural-historical activity theory as a tool of analysis

Central to the use of activity theory as a tool of analysis is the description of the elements of the activity system and a follow up with five principles of analysis as postulated by Engeström (1987). Activity theory is built from the following principles:

1. An activity system is seen as a unit of analysis.
2. An activity system "takes shape and gets transformed over time" (Engeström, 2001:137) with the analysis uncovering the history of the activity system. This is referred to as 'historicity'.
3. In an activity system, there is an understanding that the voices of the participants need to be heard. This is referred to as 'multi-voicedness'.
4. In an activity system, contradictions are driving forces of change in the activity system. Clashes between opposing points of view providing opportunities for change.
5. There is expansive transformation in activity system. A new understanding of an activity arises when participants address contradictions and question their actions, leading to new possibilities (Behrend, 2014:112).

LEDIMTALI as an activity system

In the proposal submitted to the NRF SA, gives the overall goal of LEDIMTALI as the development of high-quality mathematics teaching to enhance achievement in the high-stakes NSC Mathematics examinations of learners from low socio-economic status environments. In terms of the CHAT perspective expounded earlier, Table 2.1 presents the different components of the espoused activity system.

TABLE 2.1   Components of LEDIMTALI as an activity system

Component of activity system	LEDIMTALI	Outcome
Instruments	Different 'home-grown' documents focussed on high-quality teaching. Examples: ø narratives or part-narratives of 'actual' teaching in the classrooms of participating teachers; ø video clips of teaching aimed at the outcome; and ø designed lessons.	
Subject	The subjects are teachers, university-based experts (mathematics teacher educators), mathematicians, curriculum advisors, university-based curriculum experts and post-graduate students.	
Division of labour	ø Teachers teaching in classrooms. ø Project workers providing coaching. ø University-based experts lead the design of activities for classroom implementation, facilitate teacher learning in workshops and institutes. ø University-based experts research and produce academic artefacts for dissemination through appropriate channels such as journals, conference presentations and the theses of post-graduate students. ø University-based experts observe the teaching of lessons and produce narratives focussing on positive aspects.	Increased number of learners writing Mathematics in the NSC examinations. Increase in the quality and quantity of passes in the NSC Mathematics examination.
Community	As for 'subject' and includes learners as beneficiaries.	
Rules	ø Expertise that teachers have must be respected. ø Teachers must be respected. ø Issues that teachers bring to the table must be attended to. ø Overall trust and respect must govern interactions. ø 'Teacher blame' by university-based experts is 'out-of-bounds'. ø Have high expectations of learners. ø 'Do-ability' of distributed pedagogical ideas must underlie deliberations.	

Tools, the division of labour and LEDIMTALI

Regarding the instruments or tools, one of the approaches was to use specific-designed tools to drive deliberations about the development of teaching. By nature, these tools were discursive, and rather than only providing participating teachers with sets of ideas about high-quality teaching, one approach was to search for instances of quality and innovative teaching in the actual classroom practices of the teachers. To realise this, observations of classroom lessons were done. In addition, narratives of the lessons were produced for discussions during the CPD meetings. Appendix A is an example of a narrative of an observed lesson in trigonometry in Grade 10. The narrative thus became a tool for reflection on teaching in institutes and workshops. During these events, teachers had to consider the lesson narrative, discuss the positive features related to teaching, and present suggestions for improving the teaching. Figure 2.2 presents the responses given to the narrative in Appendix A.

Lesson evaluation: Positives

Ø Checking: Routines followed; Rules of the game.
Ø Doing on board: sharing, some learners focus on what others are doing.
Ø Monitoring: Reflective, understanding, "I must be ready," who is going to be called next; learners' errors in homework monitored.
Ø Moving as far back as Grade 8 for multiples.
Ø Connections: Trig, multiples, surds, geometry and the Cartesian plane. [All integrated.]
Ø Development of thinking skills.
Ø Correcting learners along the way (in real time).
Ø Guidance given by teacher.
Ø What, why and how questions.
Ø Invitation to get assistance if stuck.

Challenges: Openings (seeds) for improvement

Ø Little more highlighting in trigonometry.
Ø Mapping prior knowledge requirements and dealing with it before.
Ø Possibility of shifting away from the main focus.
Ø "Without using a calculator" must be stressed.
Ø Possibility of links to Grade 12 work and how it will be asked in Grade 12.
Ø Simplification 'incorrect'.
Ø Progression to surd form.

FIGURE 2.2   Responses from teachers to the teaching narrative in Appendix A

A second kind of tool is one whose design emanates from interactions during CPD activities and co-developed by all the subjects. The 'ideal' use of the tool emanates during its development, but this use is tempered at the site of implementation. In LEDIMTALI, such a tool was developed during deliberations on hampering issues teachers experience in their teaching. The neglect of homework by learners was raised. Through discussion and debate, the strategy of 'spiral revision' was arrived at. Julie describes 'spiral revision' as follows:

> Spiral revision is the repeated practicing of work previously covered. It is underpinned by the notion that through repeated practice learners will develop familiarity with solution strategies of mathematical problem types that they will come across in high-stakes examinations. Productive practicing has to do with allowing learners to develop general ways of working in school mathematics through 'deepening thinking'-like problems whilst practicing. An example of such a problem is "Factorise $ak - (k + a) + a^2$ in more than one way." The procedure suggested for implementing and sustaining spiral revision and productive practicing is that at the start of a period 2 or 3 three problems on work previously done are presented to learners. This should preferably not take more than 10 minutes of the time allocated for the lesson period. (Julie, 2013:93)

At a workshop following the development of the strategy, a teacher reported on its use in one of her classes. She brought to fore the value learners attached to the strategy expressed by a learner as "Miss, we must do more of this. I forgot the work we did in the first term." This positive instance led to the design a set of 'spiral revision' cards for work learners did in the first term. The designed cards were primarily based on problems related to the Annual National Assessments (ANA) for Grade 9. The cards were designed by the project staff with each card having about three problems and dealt with more than one topic. This is evident in Figure 2.3.

The intended implementation strategy was that teachers use the cards at the start of a lesson and give learners five to ten minutes to revise previously covered work. Teachers, however, developed their own ways of using the cards. One teacher pasted all the cards along the display boards in his classroom. He explained that he used the cards as follows: At the end of some lessons I would let each row select a card and one problem on the card which they would want to work on. Upon completion of the problem, a learner randomly selected from the row will work through the group's selected problem. He felt that this way of working "kept learners on the ball".

Grade 9   **Revision Exercises**   **Card 10**

Algebraic Expressions; Factorisation; Exponents

1. Given: $3x^2 + 6x^4 - 2x^5 + 4x^0 - x^3$
    - (a)   Is this expression a polynomial?
    - (b)   What is the degree of this expression?
    - (c)   Write this expression in descending powers of x.

2. Sally simplified this question as follows:

$$\frac{x^2 - 25}{3x} \div \frac{x - 5}{3x} = \frac{x^2 - 25}{3x} \times \frac{3x}{x - 5} = \frac{(x-5)(x+5)}{3x} \times \frac{3x}{x - 5}$$

Her answer is $(x + 5)$. Is she correct? Give a reason for your answer.

3. Is $\frac{-54}{3^2 \times 2^2} = -1$. True or False? Explain your answer.

FIGURE 2.3   A spiral revision card

Another teacher used his initiative and developed a box of work-cards for revision for the Annual National Assessment for Grade 9 Mathematics. Each card consisted of two to three problems that appeared in previous ANA examinations, as illustrated in Figure 2.4.

---

1. In diagram AC = BC. $\hat{B} = 40°$, $\hat{A}2 = 30°$

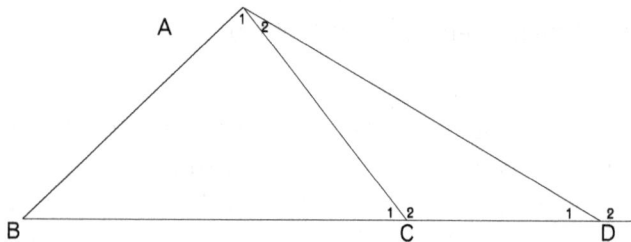

Calculate the size of D_2 with reasons

2. Simplify: $(2x - 3)(x + 1)$

FIGURE 2.4   A teacher's work-card

The teacher described how he uses the work-cards as follows: During a revision period, a learner would select any card from the box and work on the problems on the card. He [the teacher] would monitor the work and give hints and assistance based on observations of the progress of the learners' work.

The two cases illustrate that although the collectively developed tool – the spiral revision cards – was accepted, their use was adapted to comply with teachers' zones of knowledgeability and comfort in their practice. What did not change was that the adaptations for use did not disturb the focus of the outcome. This kind of tool is what we call a 'tempered tool' for classroom use collectively designed by all CPD participants. Its locus regarding the division of labour for design is all the subjects in the activity system. Its implementation and adaptations are within the ambit of the participating teachers.

The in-the-moment affordance tool is a third instrument that was 'home-grown'. It emerged during a CPD activity through observation and identification of a limited understanding of a mathematical idea. This occurred during a CPD meeting when the focus was on the developing of examination questions. One question dealt with parallel lines. Upon the feedback on the mark allocation for the question, the issue of reasons for statements emerged. During classroom observations, it came to the fore that the reason given for statements regarding the equality of angles was the name of the angle pairs rather than the lines being parallel. Figures 2.5 shows the problem and an excerpt of the mark allocation.

1.3 ABCD is a parallelogram. BE = DF. Prove that

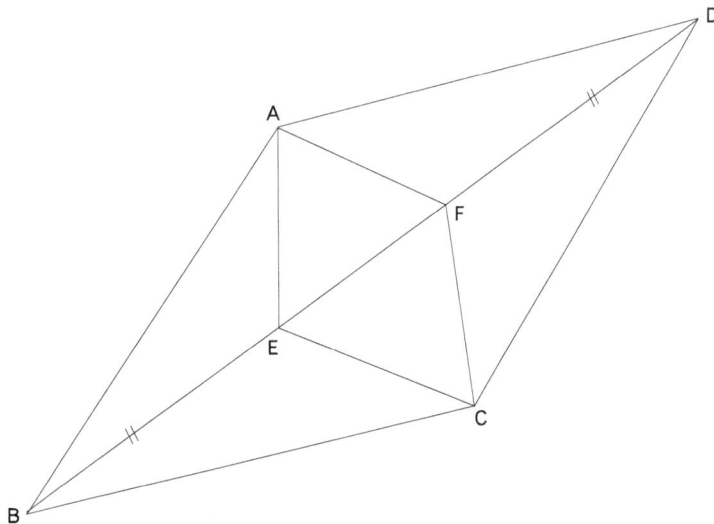

1.3.1 Δ AEB ≡ Δ CFD   (4)

1.3.2 AE ∥ CF   (3)

1.3.3 AECF is a parallelogram   (3)

FIGURE 2.5   Question and mark allocation on parallelism

During the discussion on the solution of the problem, the facilitator interrupted the teacher who was reporting and asked what he thought the learners would find difficult about the offered reason for the statement, $A\hat{B}E$ and $C\hat{D}F$. One teacher pointed out that they would name the pair of angles, alternate angles, rather than the parallelism of the lines. She continued by explaining that she addresses this issue when dealing with three intersecting lines by first developing the concepts of pairs of angles before moving to their equality under the condition of parallelism. This is linked to Euclid's fifth postulate and as Julie (2014) reports, 24% of the participating teachers indicated that they were unaware of the postulate being the underlying mathematical construct for the reason.

This notion of a tool here is akin to when a person is doing something but needs some other tool to realise a sub-task, however, the required tool is not readily available and that person ends up using whatever is available. On completion of the job, the person moves ahead with the main task, now using the appropriate tools.

The feedback on mark allocation linked to the history of actual classroom observations was thus an in-the-moment affordance tool used to strengthened teachers' knowledge of a mathematical construct. The CPD facilitator identified the reporting on mark allocation as a tool for engagement on mathematical content knowledge. On the other hand, the teacher provided an extended explanation of her way of correcting the anticipated incorrect responses learners might provide. Consequently, the division of labour was of a shared nature.

As presented in Table 2.1 on page 23 one of the instruments used was video clips of teaching aimed at the outcome. One such video clip dealt with feedback on learners' work from the assessment for learning perspective (Mathematics Assessment Resource Service, MARS, 2012). Amongst other issues, the idea of giving marks or grades was highlighted and discussed. The video illustrates and explores the provision of written comments to learners work. Although the views expressed by teachers in the video about written comments were generally positive, concern was expressed about sustaining this practice given the length of time it required.

After viewing the video clip, the LEDIMTALI participating teachers expressed similar trepidations. The challenge was how this seemingly useful strategy could be employed in LEDIMTALI. To preserve the essence of the idea of the written comments, the project staff came up with the idea that activities incorporating such comments should be developed as part of the collection of spiral revision of activities. The gist of the nature of such activities is that the responses to problems by learners and comments to the responses are provided. The task for the learners is to match the comments to the responses as shown in the following case.

A Grade 8 class was given the following task.

> Squares of different sizes are drawn and the dots inside each square are counted. Write a rule or formula to find the number of dots inside the square when you know the length of the side of the square.

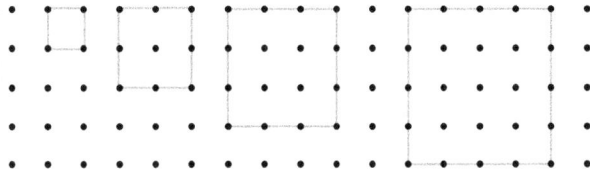

The learners handed in their work. These are the answers of three learners.

Learner A

The number of dots are square numbers. Even 0 is a square number. To get 0, 1 is subtracted from the length of the side. This gives 0 and the square number for 0 is 0.

The next number of dots $= 2 - 1 = 1$

The square number of 1 is 1.

The rule is

Number of dots $= $ (length of side)^2.

Learner B

SL	1	2	3	4	15	6
D	0	1	4	9	16	25

SL = Length of side of square.

D = Number of dots. The number of dots are all square numbers. The rule is D $= $ (SL)^2 if the SL is always the one before the next one.

So for SL $= 2$ then you square SL before the 2, which is the D $= (1)^2 = 1$ for SL.

If 1 then D $= (1)^2 - 1$ for SL. It works the same for all the other SL.

Learner C

```
0
      1
1
        3
4        5
9
```

The difference of the number of dots is the odd numbers.

Rule: Add odd numbers in order.

The next number of dots will be 16 because 7 will be added to 9.

There will be 81 dots inside if there are drawings.

The teacher did not mark their work; however, she wrote comments for the work of each learner.

Teacher's comment 1   You worked correctly by finding the difference between the number of dots. Your rule works but it will be difficult to find the number of dots inside a very big square, say one when you have 50 squares. Your rule does not connect the number of dots to the length of the side of the square. Try to find the connection.

Teacher's comment 2   The way you organised your answer by getting the number of dots in more squares is good. Your rule is correct for the example that you give. When making the rule it should work for all the information that is given but your rule does not work to get the number of dots (0) when the length of the side is 1. Check with other learners in the class how you can change your rule so that it also works for this case.

Teacher's comment 3   You obtained a correct rule. It is always wise to show that you have checked your rule with all the drawings and also try to create other drawings which show that your rule still works.

Which comment corresponds to which learner's answer?

Which comment do you agree with most? Give reasons for your answer.

The narrative above illustrates how a tool, commentaries on learners' responses, was deemed as an appropriate one by the CPD providers. Upon offering it and its operationalisation in another context via a video clip regarding the CPD tool to project teachers, the tool was found to be cumbersome and consequently would not fit seamlessly into the day-to-day teaching practices. The tool had to be adapted to become near-ready-to-use. In this instance, the locus of the division of labour was under the CPD providers.

It is not always the case that tools offered during CPD activities are explicitly used. Julie (2013) developed a teaching model to anchor pedagogical activities. To facilitate the implementation, a form (illustrated in Figure 2.6) was developed for use by teachers for preparation of lessons according to the model. Although observation of lessons revealed that pedagogical elements contained were present in the lessons (Julie, 2016), there was no visible evidence of the form being used for its designed purpose. More prominent during lesson delivery was that the textbook in use was also used for the lesson preparation. The instance demonstrates a non-useful offered tool.

Although driven by teachers' concerns, the division of labour was entirely located within the CPD providers. The use of the tool was, in a sense, rejected by the teachers due to other tools, such as textbooks, structuring their preparation of lessons. However, the CPD-providers did not entirely discard the tool. It was incorporated into an instrument to assess videos of teaching and offered to teachers to use for observation and discussion of the teaching of peers (Julie, 2016).

Intentional teaching	To teach the drawing of graphs of linear functions when the formula is given.
Mathematical objects	Concepts: graph of a linear function; x-intercept; y-intercept; gradient
	Procedure: Drawing of a graph of a linear function when the formula for the function is given; calculating the x-intercept; y-intercept and gradient of a graph of using the formula for a linear function.
Mathematical knowledge	Representation of linear relationships describe by a formula in graphic form.
Learning intention(s) (LIs)	"We are learning to" (a) draw a graph of using the formula for a linear function; (b) calculate the x-intercept; y-intercept and gradient of the graph of a linear function from the graph and the formula representing the linear function.
Success criteria (SCs)	"What I'm looking for" is that I can: (a) draw a graph of using the formula for a linear function; (b) read off the x- and y-intercept of a graph of a linear function when the graph is given; (c) calculate the x- and y-intercept when the formula for a linear function is given; (d) find the gradient of a graph of a linear function from a given graph of the function; and (e) identify the gradient of a linear function when the formula of the function is given.
Teaching and task engagement by learners	Preliminary activities (if any):
	Revise work done in Grade 8. Use tables of ordered pairs to plot points and draw graphs on the Cartesian plane.
	Lesson procedure:
	Based on the revised knowledge of Grade 8, provide learners with activities related to the LIs. Use the normal learning resources such as textbooks.
	Give learners an activity such as:

Write down the letters of the graphs representing the following linear functions:

$y = 2x + 4$; $y = -7$; $y = x$; $y = -2x - 4$

FIGURE 2.6 Lesson preparation form for Lesson 1: Grade 9, Term 3

Conclusion

A question that arises from viewing a CPD initiative through the lens of an activity system and the descriptive analysis given in the chapter is "what can we learn about tools for the provision of CPD?" Firstly, it is observable in that a variety of tools become visible during CPD work. Attention is drawn to five such meditational tools that emerged during the work of development of teaching for a particular objective. The tools are not exhaustive and for an activity system with a different object, different tools might emerge. This illustrates how tools have an uncanny capacity to evolve in work activities. During such working periods, the identification of the resources, normally embedded in activities as tools, is not in the foreground. It is by standing back later and reflecting on the activity system that the tools and their efficacy become visible.

Secondly, as CPD providers, we have learnt that for an activity system comprising subjects from different sites, it cannot be expected that all tools offered by subjects from one site will be taken on board seamlessly by the subjects from other sites. This non-acceptance is not a simple rejection. Rather, the participants from the other sites might have good reasons for rejection since their site away from the tool-offering site is in and of itself an activity system with its components. The 'new' tool offered at the continuing professional development site might compete with ones that are in a Heideggerian sense 'ready-to-hand' (Dreyfus, 1997:27). Teachers have used such ready-to-hand tools in 'a hitch-free manner' since they started teaching (Wheeler, 2013:9). Textbooks with their particular structuring for lesson presentations are examples of the ready-to-hand tools at the site of operation of where 'new' tools distributed by CPD providers must be implemented. If the offered tool is too distant from the practice of the teacher implementers, then there is a low likelihood that the tool will be used. Thus CPD providers need to have a keen sense of awareness of how 'new' tools must be presented and mediated to address this low likelihood of adoption.

Lastly, although the meditational tools are primarily offered by the CPD providers, there should be sensitivity and vigilance that teachers as participants can also facilitate the production of tools. However, care should be taken of Wittmann's advice, with teaching units viewed as tools. He warns:

> The design of substantial teaching units, and particularly of substantial curricula, is a most difficult task that must be carried out by the experts in the field. By no means can it be left to teachers, though teachers can certainly make important contributions within the framework of design provided by experts. (Wittmann, 1995:365)

Wittmann (2016:29) further suggests that "the most important service mathematics educators can render to teachers is to provide them with elaborated substantial learning environments".

Based on these considerations, we would suggest that from a CHAT perspective, CPD providers should be alert to how the various tools that are used and emerge during their interactions with teachers can be productively employed by teachers to reach the object and outcome of the teacher work-site.

References

Behrend, M.B. 2014. Engeström's activity theory as a tool to analyse online resources embedding academic literacies. *Journal of Academic Language and Learning*, 8(1):1835-5196.

Dreyfus, H. 1997. Intuitive, deliberative and calculative models of expert performance. In: C.E. Zsambok & G. Klein (eds.), *Naturalistic decision making*. Mahwah, NJ: Lawrence Erlbaum. pp.17-28.

Engeström, Y. 1987. *Learning by expanding: an activity-theoretical approach to developmental research*. Helsinki: Orienta-Konsultit.

Engeström, Y. 1993. Developmental studies of work as a test bench of activity theory: The case of primary care medical practice. In: S. Chaiklin & J. Lave (eds.), *Understanding Practice: Perspectives on activity and context*. Cambridge, UK: Cambridge University Press. pp.64-99. https://doi.org/10.1017/CBO9780511625510.004

Engeström, Y. 1999. Innovative learning in work teams: Analysing cycles of knowledge creation in practice. In: Y. Engeström, R. Miettinen & R. Punamäke (eds.), *Perspectives on activity theory*. Cambridge, UK: Cambridge University Press. pp.377-404. https://doi.org/10.1017/CBO9780511812774.025

Engeström, Y. 2001. Expansive learning at work: Towards an activity theoretical re-conceptualisation. *Journal of Education and Work*, 14(1):133-156. https://doi.org/10.1080/13639080020028747

Engeström, Y. 2010. Activity theory and learning at work. In: M. Malloch, L. Cairns, K. Evans & B.N. O'Connor, *The Sage handbook of workplace learning* (Ch. 7). Helsinki: University of Helsinki.

Goodchild, S. 1997. *An exploratory study of year ten students' goals in the mathematics classroom*. DPhil thesis, University of Exeter.

Julie, C. 2013. Towards a model for intentional teaching for improving achievement in high-stakes mathematics examinations. In: Z. Davis & S. Jaffer (eds.), *Proceedings of the 19th Annual Congress of the Association for Mathematics Education of South Africa* (Vol. 1). Cape Town: AMESA. pp.86-96.

Julie, C. 2014. Emergences and affordances as opportunities to develop teachers' mathematical content knowledge. *International Journal of Mathematical Education in Science and Technology*, 45(3):428-437. https://doi.org/10.1080/0020739X.2013.851809

Julie, C. 2016. *Does a CPD initiative focusing on the development of teaching to enhance achievement outcomes in high-stakes mathematics examinations work?* A report on the first 5-year phase of the Local Evidence-Driven Improvement of Mathematics Teaching and Learning Initiative – LEDIMTALI. Bellville, South Africa: University of the Western Cape.

Leont'ev, A.N. 1981. *Problems of the development of the mind.* Moscow: Progress Books.

MARS (Mathematics Assessment Resource Service). 2012. How can I respond to students in ways that improve their learning? *Professional development module: Formative assessment.* Nottingham: MARS, Shell Centre, University of Nottingham.

McNicholl, J. & Blake, A. 2013. Transforming teacher education, an activity theory analysis. *Journal of Education for Teaching*, 39(3):281-300. https://doi.org/10.1080/02607476.2013.799846

Roth, W-M. & Tobin, K. 2002. Redesigning an 'Urban' Teacher Education Program: An activity theory perspective. *Mind, Culture, and Activity*, 9(2):108-131. https://doi.org/10.1080/02607476.2013.799846

Vygotsky, L.S. 1978. *Mind in society: The development of higher psychological processes.* Cambridge, MA: Harvard University Press.

Wake, G.; Swan, M. & Foster, C. 2015. Professional learning through the collaborative design of problem-solving lessons. *Journal of Mathematics Teacher Education*, 19(2):243-260. https://doi.org/10.1080/02607476.2013.799846

Wheeler, M. 2013. Martin Heidegger. In: E.N. Zalta (ed.), *The Stanford encyclopedia of philosophy* (Spring, 2013 Edition). Available: https://stanford.io/2ZxHxxB (Accessed 21 August 2014).

Wilson, V. 2014. Examining teacher education through cultural-historical activity theory. *Teacher Education Advancement Network (TEAN) Journal*, 6(1):20-29.

Wittmann, E.C. 1995. Mathematics education as a 'design science'. *Educational Studies in Mathematics*, 29(4):355-374. https://doi.org/10.1007/BF01273911

Wittmann, E.C. 2016. Collective teaching experiments: Organizing a systemic cooperation between reflective researchers and reflective teachers in mathematics education. In: M. Nührenbörger, B. Rösken-Winter, C-I. Fung, R. Schwarzkopf, E.C. Wittmann, K. Akinwunmi & F. Schacht (eds.), *Design science and its importance in the German mathematics educational discussion.* Dordrecht: Springer (Open). pp.26-34.

APPENDIX A

Grade 10

Trigonometry was the topic. The teacher informed me that they did special angles. The teacher did the problem (a) If sin a = $\frac{4}{3}$ find cot A + tan A. Learners had to do (b) If sin A= $\frac{3}{8}$, find cosec A – sec A and (c) If 13 cos A = 12, find sec A + sin A. [Teacher informed that SOHCAHTOA was given to learners as 'cue' for the trigonometric ratios.]

Two learners worked on the solutions of (b) and (c) on the board whilst the teacher monitored the completion of homework of the other learners.

Learners' work on board

b)

c)

$x^2 = 8^2 - 3^2$

$x^2 = 64 - 9$

$x^2 = 55$

$\frac{8}{3} - \frac{8}{\sqrt{55}}$

$\frac{8\sqrt{55} - 24}{3\sqrt{55}}$

$13^2 = 12^2 + y^2$

$13^2 - 12^2 = y^2$

$169 - 144 = y^2$

$25 = y^2$

$\sqrt{25} = \sqrt{y^2}$

$5 = y$

sec A + sin A

$\frac{5}{13} + \frac{5}{12} = \frac{169 + 60}{156} = \frac{229}{156}$

Attention is directed to (b).

Teacher   Is $\sqrt{55}$ a perfect square or non-square?

Class   Non-square.

Teacher   Should the answer be left in surd form? [A learner explains how (b) was done.]

Teacher   Angle A is acute. In which quadrant is it? Which quadrant?

Class   First quadrant.

Teacher explains and represents angle A and the information in the Cartesian plane (given in the accompanying diagram.) SOHCAHTOA is used to get the sides.

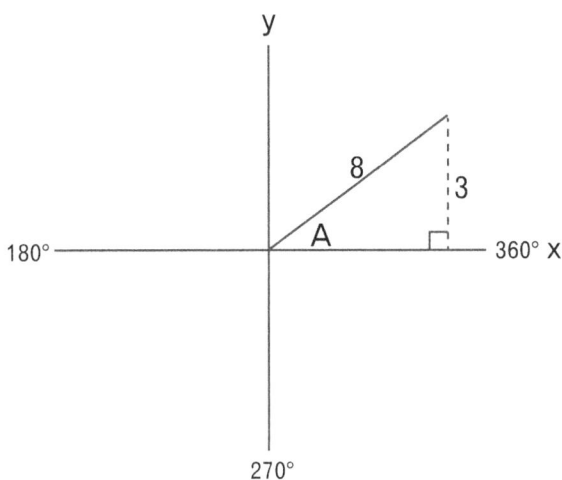

Teacher   Which theorem was used to get the adjacent side?
Class   Pythagoras.

The focus shifts to the final presented answer for (b).

Teacher   Is the LCD correct? Did she simplify correctly?

The teacher demonstrates the correctness.

Teacher   What happens if $\dfrac{8\sqrt{55}-24}{3\sqrt{55}}$? Are we going to leave it just like that? What are we going to do? Stop there and say "This is the answer"? I am not saying that the problem's answer is incorrect. Just getting you to understand the simplification. Which will you simplify? Numerator? Denominator?

Class	Numerator.
Teacher	Is it not the same as $2x - 4x$?
Class	$-2x$
Teacher	How will you simplify this [the numerator of the surd expression.]? $8\sqrt{55} - 24\sqrt{55}$ is the same [referring to $2x - 4x$]. $\dfrac{-16\sqrt{55}}{3\sqrt{55}}$ is arrived at.
Teacher	Final answer? Is this [referring to the fraction in the above line] our answer?

The majority of learners say 'no', but 'yes' is also heard. One learner wanted to know about 'plus times minus'. This was left unanswered. $\sqrt{55}$ was 'cancelled' and $\dfrac{-16\sqrt{55}}{3\sqrt{55}} = \dfrac{-16}{5}$ was left on the board.

Teacher	I am checking just whether you can still do simplification.

The class is requested to read (c) aloud which they do.

Teacher	What is the difference between (b) and (c)? 'c' has no fraction.

The learner, who did the problem on the board, explained how she went about it including in her explanation the Cartesian plane representation. The given (13 cos A = 12) is recalled.

Teacher	How did you calculate [referring to the unknown side] that side? 12 is the adjacent side? Why?
Learner	Divide by 13. We are looking for cos A, not 13 cos A.
Teacher	It is the same as $2x = 2$. $\dfrac{2x}{2} = \dfrac{2}{2}$ What you do to the RHS you do to the LHS. [It is] the same with 13 cos A. Multiplication 13.cos A. Divide by what you are not must not be the coefficient.] $\cos A = \dfrac{12}{13} \left(\dfrac{Adj}{Hyp} \right)$ is written on the board by using SOHCAHTOA to get learners to define cos. The sketch for the Cartesian representation is also drawn.

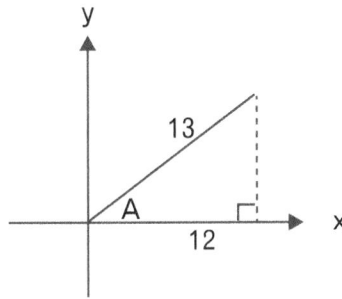

During the construction, the learners were answering teacher-directed questions.

Teacher	Why? [referring to the unknown side].
Class	y-axis
Teacher	Positive or negative? 5 correct. We know all our sides. Sec A is the reciprocal of: $cos\ A = \dfrac{Adj}{Hyp}$, $sec\ A = \dfrac{Hyp}{Adj}$ The $\dfrac{5}{12}$ that was written for sec A is corrected to $\dfrac{13}{12}$. The answer for (c), $\dfrac{169 + 60}{156}$ was, however, deemed correct.
Teacher	LCD or LCM. How did you get 156? Did you use your calculator?
Learner	Used my calculator for 12 times 13.
Teacher	Did you use your calculator directly? $= \dfrac{13}{12} + \dfrac{169 + 60}{156} + \dfrac{5}{3}$? Now we are going to do it the long, long way.
Teacher	Multiples of 12 and writes with the multiples given by the learners.
M12	12, 24, 36, 48, 60, 72, 84, 96, 108, 120, 132, 144, 156, [Although 156 was reached the teacher said that they are going to stop at 144 to see what is going to happen.] A similar procedure is followed for the multiples of 13.

M13

13, 26, 39, 52, 65, 78, 91, 104, 117, 130, 143, 156 [After 78 there was a slight pause and the teacher wanted to know what they must do to which a learner responded: 'Add 13'.]

Teacher reverts to M12 and says 'So far do we have lowest one? [Reference is to M12 presently ending at 144]. We must continue. What now, 156?'

The learners now counted the multiples up to $\frac{229}{156}$ for M12 and M13 and are arrived at. This shows the learners that for the method used to find the LCM they need to think 'about the time' this method takes.

Teacher (referring to $\frac{229}{156}$)

Common factor? How do we know there is a common factor?

The learners are instructed to correct their work and call on the teacher when they need assistance.

| O3 |

ASSESSMENT OF TEACHERS' MATHEMATICAL CONTENT KNOWLEDGE THROUGH LARGE-SCALE TESTS: WHAT ARE THE IMPLICATIONS FOR CPD?

Cyril Julie

Introduction

In South Africa, particularly in 2013, the media placed much emphasis on teachers' mathematical content knowledge. The reports in the popular media arose from various diagnostic measures that were used to come to this conclusion, and the measurements – quantitative or qualitative – primarily focused on performances by learners and teachers of these learners in school mathematics. A conclusion drawn from the performances in these tests was that teachers' content knowledge of school mathematics is lacking. This deficiency of mathematical content knowledge of teachers is deemed as a major contributor to the low achievement in mathematics. This is indicated in the Trends in Mathematics and Science Study (TIMSS), in public schools (see Reddy, Visser, Winnaar, Arends, Juan, Prinsloo & Isdale, 2016).

This chapter focuses on the items used in three research studies related to the measurement of teachers' mathematical content knowledge. The structuring effects of the test-taking conditions are also delved into. The purpose of the analysis was to ascertain whether these items met the criteria of acceptable instruments to reach the conclusion regarding teachers' lack of mathematical content. It is done in the spirit proposed by Hans Freudenthal, namely that a researcher should indicate "whether and to which degree he (sic) has checked the research he cited or quoted" (Freudenthal, 1979:303). For the purpose of the research reported in this chapter, the checking referred to by Freudenthal is applied to items comprising tests assessing teachers' mathematical content knowledge.

Selection of studies and analysis framework

The criteria used for selecting the studies were that they had to be large-scale, not opportunistic. Also, the population of interest from which the sample was drawn had to be clear and the data had to be analysed quantitatively. To date, only three such studies investigating South African teachers' mathematical content knowledge were conducted.

Taylor (2011) and Taylor, Van der Berg and Mabogoane (2013) report on the National School Effectiveness Study (NSES) where schools were randomly sampled from all provinces except Gauteng. The study was longitudinal, and part of the study was a five-item test for Grades 3 and 5 teachers from the sampled classes of the selected schools. The test was an open-response one with the respondents having to construct his or her responses to the items. Taylor et al. do not specifically indicate how many teachers wrote the mathematics teacher test, but from a table, it was calculated that 236 teachers took the test (Taylor et al., 2013:71).

A study reported by Carnoy, Chisholm and Chilisa (2012) was confined to the North-West Province. It was part of a comparative study between the province and the south-eastern region of Botswana and focused on Grade 6 teachers. The teacher test had 24 items with 11 items having multiple sections which gave a total of 63 items to which teachers had to respond. Sixty-one teachers in the North-West Province completed the test.

The test scores analysed by Spaull (2013) and Venkatakrishnan and Spaull (n.d.) were those in the 2007 survey of the Southern and East African Consortium for Monitoring Educational Quality (SACMEQ III) to ascertain the mathematics content knowledge of Grade 6 mathematics teachers. The test had 42 multiple-choice items and "498 Grade 6 mathematics teachers [were] included in the SACMEQ 2007 South Africa survey and that of these, 401 (81%) wrote the mathematics teacher test" (Venkatakrishnan & Spaull, n.d.:11). Venkatakrishnan and Spaull re-analysed the SACMEQ III teacher test data in terms of mathematical content strands:

1. Number and operations,

2. Fractions, decimals, and proportional reasoning,

3. Patterns, graphical reasoning and algebra, and

4. Shape and space (Venkatakrishnan & Spaull, n.d.:12).

The criterion they used to ascertain whether a teacher had command of the requisite knowledge at a particular grade level was 60% or higher for the items comprising the strand. With this approach, they found "that the vast majority (79%) of South African Grade 6 mathematics teachers were classified as having content knowledge levels below Grade 6" (Venkatakrishnan & Spaull, n.d.:14).

Both the SACMEQ III test and the one used in the study of Carnoy et al. (2012) were multiple-choice closed response ones. It is worth noting that Taylor and Taylor (2013) eventually focus their discussions and conclusions on teacher performance on the SACMEQ III survey.

The analysis reported in this chapter focuses on the items reported in the three studies since these were the items used in the reports to support the argument made about teachers' mathematics content knowledge. A set of constructs was used to conduct the analysis. The constructs are (a) ambiguity of item formulation from test construction literature, (b) the retention problem – inability to recall information due to non-use of knowledge that was acquired previously, (c) knowledge valued and legitimated through boundary objects in a practice, (d) disciplinary fidelity of the items with respect to mathematics and mathematics education, and (e) openness or not of items with respect to system 1 cognitive processing. These constructs are interrelated and overlapping. The analysis indicates those items that fall into more than one of these categories, these are indicated.

Before providing the analysis, the contexts within which the teacher testing took place is provided, as in some of the studies it is explicitly stated that the reports should be read by keeping the contexts in mind.

Context of the teacher test-taking

Tests are approached with different degrees of seriousness and effort by test-takers. The highest levels of seriousness and effort to perform well are exerted when tests are of a high-stakes nature. High-stakes tests are ones with consequences of importance to the test-takers. The test given in the study by Carnoy et al. (2012) was part of a larger questionnaire the teachers had to complete, and they caution against over-interpretation of the obtained test scores. They stipulate that the results should be read in conjunction with:

> [teachers] not taking the questionnaire very seriously (it had absolutely no consequences for them), or after many years (the majority of teachers were over 40 years old) they had forgotten much of the mathematics they had learnt or had not had a good initial base of mathematics training … the teacher information part that preceded the test part of the questionnaire was long and asked a lot of detailed information on their teacher training and opinions on various education issues. Thus, by the time they got to the test section, they may have been short of time or just tired of answering questions. (Carnoy et al., 2012:94)

In addition to mentioning fatigue; the seriousness with which teachers took the test because of its non-high-stakes nature and non-retention of content due to the time lag between the test and when teachers studied appropriate mathematics during their initial teacher training (teachers had an average age of ±40 year) (also note that "only about one out of five named mathematics as their primary subject of specialisation during their teacher education" (Carnoy et al., 2012:92). Venkatakrishnan and Spaull (n.d.:11) also allude to teachers' willingness to participate and found "large discrepancies between teacher response rates by province". This is ascribed to competent teachers' refusal to do the test and is related to the apparent non-serious disposition with which teachers approach tests of this nature. Taylor and Taylor (2013:227) suggest that a reason for the non-participation of about 15% of the selected teachers that the "teachers with poor language and/or mathematical proficiency [declined] to write".

From another perspective, teachers who were called to but did not participate were simply exercising their right to voluntary participation which is a crucial element for studies of this nature. It is common knowledge that teachers are more and more exercising their right to voluntary participation in educational research matters relating to their practice. It is for this very reason that many projects provide incentives for teachers to participate in research endeavours. These incentives might be in cash as stated in a project proposal to the National Science Foundation in the United States of America where the proposers offered "an incentive ($100) for participating in each round of data collection" to participating teachers. Another form of incentive is to offer participating teachers credits towards higher degrees for participating in a research-driven development initiative. This last-mentioned was relayed to me from by a visiting mathematics educator who directed a major project on the use of ICT in mathematics teaching.

This brings to the fore that the seriousness and willingness of participation in research endeavours that have no direct consequences (or benefits) for teachers cannot be summarily dismissed. Seriousness and willingness to participate in tests cannot be written off as inconsequential and can have a contaminating effect on the authenticity of the data that are subjected to analysis.

A final issue related to the contexts of the test related to the test itself is provided by a statement by Taylor et al. that declares that their test "proved to be too short to provide either a rich picture of teacher knowledge or satisfactory associations in the modelling exercise [related to teachers' subject matter knowledge and learner achievement]" (Taylor et al., 2013:19).

All three studies caution against over-interpretation of their results because of the various issues related to the context within which teachers engaged with the tests, and in one case, the length of the test.

Analysis of the available items in the selected studies

This section presents an analysis of the available items of the selected studies in terms of the constructs given above.

Ambiguous formulation of items

Literature abounds on the criteria test items should fulfil to be considered 'good' items. Mathematics objective-type items are the favoured type because they leave "no room for doubt about what is a fully correct answer" (Wynne-Wilson, 1978:182). This implies that the items must be unambiguously formulated so that there is no uncertainty as to what and how the test-taker has to respond to the item.

The test presented by Taylor (2011:5) had the item "10 days 75 hours can be written as … days … hours". This item is ambiguous because it is prone to many correct responses such as, for example, 13 days 3 hours, 12 days 27 hours, and 10 days 75 hours. It is not reported which response was taken as the correct one. Another ambiguous item is the one given in Figure 3.1. Carnoy et al. (2012) present this an item as dealing with teachers' pedagogical content knowledge.

Aneen says that $2\frac{2}{3}$ equals $\frac{4}{5}$ and she uses the figure below to demonstrate her assertion. Why is her reasoning not correct?

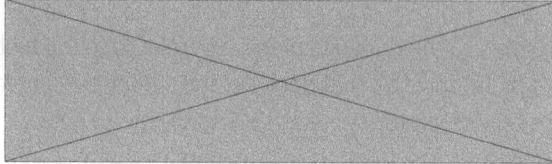

$2\frac{2}{3}$ puts block together $\frac{4}{5}$

Choose the BEST explanation why Aneen's reasoning is not correct.

(Mark ONE answer)

A.   Because $2\frac{2}{3}$ is not $\frac{4}{5}$ because it is $\frac{8}{3}$.

B.   Because she is using different units to represent the whole part and the proper fraction in $2\frac{2}{3}$.

FIGURE 3.1   Test item on pedagogical knowledge

Source   *Carnoy, et al. (2012:97)*

It is obvious that the 'correctness' of the response to the item hinges around the interpretation of 'BEST'. Carnoy, et al. (2012) report that the Grade 6 teachers scored low on the item. This implies that the test designers had their interpretation of 'BEST' and assigned the mark for correctness according to their interpretation. A spread of the responses across the four response categories is not provided and given the procedural culture of teaching the conversion of mixed numbers to improper fractions it is highly likely that teachers would favour 'A'. Also given that only one response had to be chosen, teachers could have been bamboozled. This is because under non-testing conditions they would first obtain the correct response and through a backward-chaining method work towards a pictorial representation of the situation.

Similar ambiguities of item formulation were also found in TIMSS mathematics items (Wang, 2011).

The retention problem – inability to recall information due to non-use of knowledge that was acquired previously

Carnoy et al. are explicit about this category. They assert that the purpose of their test was attempting "to assess how much knowledge teachers had retained and were bringing to their teaching" (Carnoy et al., 2012:93-94). The retention is related to the mathematics courses they had studied during their initial teacher education. Fowler (1995) points to this phenomenon in test-construction. He suggests that in test-construction there should be guarding against "questions to which [respondents knew] the information at some point but being unable to recall the information accurately and in detail required by the question" (Fowler, 1995:9). This issue is related to retention and forgetting due to non-use of knowledge. Carnoy et al. admit this when they state that teachers have "forgotten much of the mathematics they had learnt … since most [teachers] have not done mathematics as a subject for many years, it is not surprising that on average they fared relatively poorly on the teacher mathematics knowledge questionnaire" (Carnoy et al., 2012;94-102). An item related to this issue is the one given in Figure 3.2. They found that teachers found this item 'more difficult' (Carnoy et al., 2012:97) and they correctly identified it as an item requiring "higher-level understanding of rational numbers." The 'higher-level understanding' is linked to rational numbers being algebraic numbers, cardinality and countability of sets of rational numbers.

My knowledge and experience of mathematics curricula for initial teacher education programmes for teachers to teach mathematics in primary schools in South Africa is that content of this nature is not part of such curricula. There is, however, a possibility that teachers might have encountered that the set of rational numbers are countably infinite through some watered-down version akin to "the number of fractions between two numbers is infinite". The likelihood is small that this mathematical idea would have been taken to a deeper mathematical level and given further absence of it in textbooks and curriculum documents it is highly plausible that teachers have forgotten it.

How many decimal numbers are there between 0,30 and 0,40? (Mark ONE answer)

A.   9
B.   10
C.   99
D.   An infinite number

FIGURE 3.2   Test item related to retention
Source   *Carnoy et al. (2012:97)*

Knowledge valued and legitimated through boundary objects in a practice

The non-use of previously acquired knowledge is closely linked to the knowledge valued and legitimated through the boundary objects in use in practice. Boundary objects are "those … objects which inhabit several intersecting social worlds and satisfy the informational requirements of each of them" (Star & Griesemer, 1989:393). In the school mathematics enterprise, "boundary objects include the curriculum materials, the State Standards, and reports of students' test scores [and they are] reifying object[s] [which are] relatively transparent carriers of meaning for members of the community in which it was created" (McClain & Cobb, 2004:286). Teachers' encounters with school mathematics are primarily through these boundary objects. There is, therefore, a good likelihood that if items for testing teachers' mathematics content knowledge are not consonant with those constituted by the knowledge distributed by the boundary objects, then their correct responses will be low.

A popular item used to gauge teachers' content knowledge is the order of operations. What should be borne in mind is that the order of operations is a mathematical convention, different from mathematical concepts, procedures or proofs as objects of mathematics. Mathematical conventions are agreements by mathematical workers and there exist no methods to logically construct them from other mathematical entities. (As an aside, it should be noted that there is a mathematical convention where operations are performed sequentially from left-to-right. This is the Reverse Polish Notation (RPN) where push numbers and stacking are used to allow calculating starting with the numbers from the left. For $2 + 3 \times 2$ the calculation notation in RPN will be $2\ 3\ 2 \times +$ which through push numbers and stacking will give 8 as the answer.)

In some sense, mathematical conventions are things to be remembered and recalled. Historically, with respect to school mathematics, the convention of the order of operations is normally first encountered with the operations of ordinary fractions and the mnemonic that came into existence to remember it is BODMAS (Brackets, Of, Division, Multiplication, Addition, Subtraction).

The teacher mathematics test of Taylor (2011) does not contain an item on the order of operations and Taylor and Taylor (2013) use the 2007 SACMEQ survey to support their arguments. Figures 3.3 and 3.4 show the items related to the order of operations instruments used by Carnoy et al. (2012) and Venkatakrishnan and Spaull (n.d.).

Ms Bahadur asked her learners to write expressions that, when evaluated, give an answer of 10. After reviewing her learners' work, she was pleased that many of her learners used more than one operation in their expressions. However, she was concerned that their notation was not mathematically correct. Determine, if each expression as written below, equals 10.

(Circle 1 or 2 to indicate EQUALS 10 or DOES NOT EQUAL 10 for each expression.)

Expression	Equals 10	Does not equal 10
A. $2 + 3 \times 2$	1	2
B. $8 + 9 - 5 + 2$	1	2
C. $15 - 2 - 3$	1	2
D. $80 \div 4 \times 2$	1	2
E. $6 + 2 \times 2$	1	2

FIGURE 3.3   Order of operations item

Source   *Carnoy et al. (2012:96)*

QUESTION 6

$10 \times 2 + (6 - 4) \div 2 =$

A.   11
B.   12
C.   20
D.   21

FIGURE 3.4   Order of operations item

Source   *Venkatakrishnan and Spaull (n.d.:24)*

The item of Carnoy et al. requires teachers to effectively assess whether the responses provided by learners were correct. They assert that "teachers did reasonably well ... on parts C and D of this item" and their performance was 'much lower' on A, B and E. (Carnoy et al., 2012:16). The magnitude of the much lower performance is not explicitly provided. Responses to A and E are linked to multiplication being executed before addition according to the convention for the order of operations. B is linked to "subtraction does not follow addition when parentheses or brackets are absent from the mathematical sentence" (Carnoy et al., 2012:96). This phrase is not understandable since when brackets are absent it does not matter whether addition is done before subtraction and vice-versa since addition of integers is commutative. Venkatakrishnan and Spaull (n.d.) assert their item is related to the order of operations given in the CAPS document as "multiple operations on whole numbers with or without brackets" (DBE, 2011:14).

An examination of the CAPS document (DBE, 2011) and Grade 6 textbooks in use, such as the one by Bowie, Gleeson-Baird, Jones, Morgan, Morrison et al. (2010) reveal that the use of all four basic operations at the same time in a stand-alone problem as the above items are rarely, if ever, used. I could not find any problem in these documents where it is used as such. In these boundary objects, the convention of the order of operations is not used in and of itself. It is used in conjunction with other mathematical ideas, such as the decomposition of numbers and the distributive law. This is understandable given the importance of the convention of the order of operations in later school mathematics topics such as algebra.

Taylor and Taylor (2013) concur with this use of the convention for the order of operations. However, the test items in Figures 3.3 and 3.4 did not focus on teachers' knowledge about the use of the convention for number decomposition and the distributive law which is part of the curriculum they teach. I contend that the low scores teachers obtained for these items are more a function of the absence of items of this nature as components of the boundary objects rather than an indicator of inadequate teacher content knowledge of the order of operations.

As a final issue around the order of operations, Wu (2007) believes that in most instances questions related to the order of operations are of a 'trap' kind. He uses the problem "Evaluate $4 + 5 \times 6 \div 10$" to illustrate this point and states:

> Now one never gets a computation of this type in real life, for several reasons. In mathematics, the division symbol $\div$ basically disappears after Grade 7. Once fractions are taught, it is almost automatic that $6 \div 10$ would be replaced by $6 \times \frac{1}{10}$. Moreover, if anyone wants you to compute $4 + 5 \times 6 \div 10$, he would certainly make sure that you do what he wants done and would put parentheses around $4 + (5 \times 6 \div 10)$ for emphasis. In a realistic context then, $4 + 5 \times 6 \div 10$ would have appeared either as $4 + (5 \times 6 \div 10)$, or $4 + (5 \times 6 \times \frac{1}{10})$. The original problem is, therefore, a kind of Gotcha! parlor [sic] game designed to trap an unsuspecting person by phrasing it in terms of a set of unreasonably convoluted rules. (Wu, 2007:6)

Disciplinary fidelity

In the construction of mathematical models, normally the derived models are tested for their fidelity. This is the "the preciseness of a model's representation of reality" (Giordano & Weir, 1985:33). In the same sense, one can view the disciplinary fidelity of items in a test as its preciseness in terms of the constructs constituting the discipline. Items testing teachers' mathematical content knowledge should thus conform to the corpus of accepted ideas of the discipline. Items with low mathematical fidelity are normally open to multiple correct representations. With advances in mathematics education, there is also a body of knowledge related to forms of presentation and representation of mathematical statements that make them prone for learners developing misconceptions of mathematical entities. Accordingly, test items that do not pay sufficient attention to avoiding the cementing of such misconceptions are of low fidelity concerning mathematics education.

The items in Figures 3.5 and 3.6 have low mathematical fidelity.

$\frac{x}{2} > 7$ is equivalent to:

A. $x > 14$
B. $x < 14$
C. $x > 5$
D. $x < \frac{7}{2}$

FIGURE 3.5   Algebraic inequality item

Source   *Venkatakrishnan and Spaull (n.d:25)*

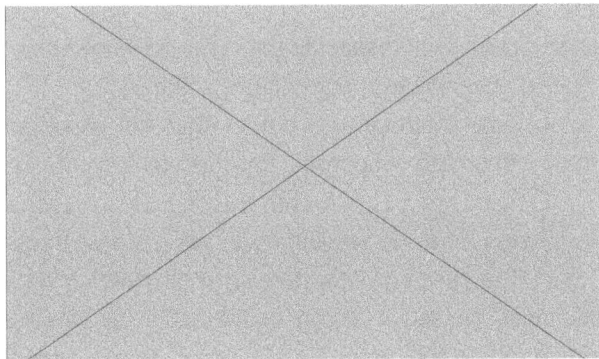

FIGURE 3.6   Perimeter item

Source   *Taylor (2011:5)*

For the inequality in Figure 3.5, the respondents have to assume that the domain is the set of positive real numbers. Without this assumption, there is no correct response category since the variable, x can be positive or negative. Similarly, for the five-sided polygon in Figure 6, the item requires that an assumption about right-angles or parallelism be made for a unique response to be obtained. This will allow the figure to be dissected into rectangular shapes from which the lengths of the unknown segments can be found. From the information provided, many different closed five-sided figures can be constructed. I surmise that the constructors or the test expected the teachers to 'see' that either of these assumptions hold due to the visual appearance of the figure. The follies of this kind of expected 'seeing' are well known. This is clearly shown in Figure 3.7 where what is 'seen' are segments of different lengths. Hence for a correct response, the assumption must be clearly stated, otherwise, as tests go, full marks cannot be awarded. I had no access to the schedule of marking and hence have to assume that the 31% and 30% of Grades 4 and 5 teachers respectively who answered correctly indicated the assumption.

FIGURE 3.7   'Seeing' equal lengths as different

Low disciplinary fidelity with respect to the corpus of knowledge in mathematics education is also observable in some items. Item 3, " $\frac{1}{4} + \frac{3}{5}$ = ___ " (Taylor, 2011:5), has no clear instruction. I gather that the expected answer is $\frac{17}{20}$. There are many alternate responses for this item as it stands. Some are $\frac{1}{4} + \frac{3}{5} + \frac{2}{5} : \frac{1}{4} + \frac{30}{50}$ and $\frac{1}{4} + \frac{30}{50}$. However, the test-taking teachers have to view the equality (=) sign as a "do something to get a single fraction on the right-hand side" cue. The research literature in mathematics education abounds with the difficulties caused in doing mathematics by viewing (and fostering the understanding of) the equality relation in this manner.[1] Although the item in Figure 3.4 is multiple-choice in nature, it also fosters this erroneous

---

[1]   See, for example, Essien and Setati (2006).

use of the equality sign. With the equality sign viewed as a reflexive, symmetric and transitive relation some other possible correct responses are: $10 \times 2 + (6 - 4) \div 2$; $20 + 2 \div 2$; $10 \times 2 + 6 \div 2 - 4 \div 2$.

Wang (2011) addresses a similar disciplinary fidelity issue with a TIMSS mathematics item stated as "the sum of the interior angles of a polygon with n sides = ___ 180°. "He states the "grading rubrics indicated that the underline space can be filled out by 'n-2 with or without brackets.'" Although the brackets did not seem to matter in the rubrics, they make a meaningful difference in mathematics.

Openness of items with respect to System 1 cognitive processing

Shane Frederick proposes the notion of two systems of metacognitive difficulties in analytic reasoning when people deal with particular kinds of problems. These are "System 1 processes [which] occur spontaneously and do not require or consume much attention [and] System 2 processes – mental operations requiring effort, motivation, concentration, and the execution of learned rules" (Frederick, 2005:26). For particular formats and formulations of problems, intuitive solutions come to mind quickly because of the operation of System 1 processes, particularly with respect to not paying attention to all the salient features of the problem. Frederick (2005) subjected 3 428, mostly undergraduate, students in the USA to a three-item test (see Figure 3.8 containing tasks to ascertain whether the cognitive processing was of the System 1 or system 2 kind. Some students were from elite institutions such as Harvard, Massachusetts Institute of Technology, Princeton and the University of Michigan.

1. A bat and a ball cost $1.10 in total. The bat costs $1.00 more than the ball. How much does the ball cost? _____cents
2. If it takes 5 machines, 5 minutes to make 5 widgets, how long would it take 100 machines to make 100 widgets? _____minutes
3. In a lake, there is a patch of lily pads. Every day, the patch doubles in size. If it takes 48 days for the patch to cover the entire lake, how long would it take for the patch to cover half of the lake? _____days

FIGURE 3.8   Frederick's cognitive reasoning test
Source   Frederick (2005:27)

The marks awarded to the test were: 0 – none of the items correct; 1 – one item correct; 2 – two items and 3 – all items correct. Sixty-one percent of the respondents either scored 0 or 1. Of the incorrect answers, "the posited intuitive answers (10, 100 and 24)" (Frederick, 2005:27) demonstrate that focusing on the surface level features of the problem generates the incorrect responses for the problems types.

Adam, Oppenheimer, Epley and Eyre (2007) did a pilot study with 40 undergraduate students of Princeton University to ascertain if a change in some surface-level feature of Frederick's (2005) test would affect students' performance. The 40 students were randomly assigned to complete the test in an easy-to-read black Myriad Web 12-point font and difficult-to-read 10% grey italicised Myriad Web 10-point font. They label the 12-point font version as the fluent font condition and the other one the disfluent font condition. They found that

> participants answered more items ... correctly in the disfluent font condition ... than in the fluent font condition. Whereas 90% of participants in the fluent condition answered at least one question incorrectly, only 35% did so in the disfluent condition [and] Finally, participants in the fluent condition provided the incorrect and intuitive response more often (23% of responses) than did participants in the disfluent condition (10% of responses). (Adam et al., 2007:570)

The aforementioned studies imply that the production of incorrect responses is not a simple function of whether or not respondents know mathematics but rather that the numbers used in these problems make them prone for the generation of incorrect responses due to System 1 cognitive processing.

The item in Figure 3.9 given to teachers is of a similar kind as those in Frederick's (2005) cognitive reasoning test.

QUESTION 9

If the height of a fence is raised from 60 cm to 75 cm, what is the percentage increase in height?

A. 15%
B. 20%
C. 25%
D. 30%

FIGURE 3.9   SACMEQ III test item similar to Frederick's cognitive reasoning item

Source   *Spaull (2013:26)*

The research by Frederick (2005) and Adam et al. (2007) points in the direction of the incorrect responses to the above being produced by System 1 processes. Changing a feature or features of the item by forcing attention might lead to different, normally correct, responses.

Given the proximity of the item to those of Frederick, the question begs whether the item is a mathematics problem, a puzzle, a cognitive reasoning task used in psychological testing or something else. Spaull (2013:26) views it as an "application of percentage calculations" problem. It is accordingly perceived as a problem where mathematics has to be applied to a real-life situation. Pollak (1969) classifies applications-like problems of this nature as mathematical problems dressed up as real-life situations. A problem of this kind is rarely, if ever, found in real-life. In the practice of raising fences, one will find it hard to find workers speaking about "percentage increase in height". My own experience of observing the practice raising vibracrete fences rendered that the talk is around "met hoeveel slabs moet os die fence raise?" (With how many slabs must we raise the fence?). That the question formulation and the heights, 60 cm and 75 cm, used had a marked influence on the responses is further supported by the percentage of correct responses for another item, linked to percentages, of the SACMEQ III mathematics teacher test given in Figure 3.10.

QUESTION 23

Last year there were 1 172 students at Griffin High School. This year there are 20% more students than last year. Approximately how many students are at Griffin High School this year?

A. 1 200
B. 1 400
C. 1 600
D. 1 800

FIGURE 3.10   Percentage item

Source   Venkatakrishnan and Spaull (n.d:24)

This question was classified as a Grade 8 level question and the average percentage teacher score was 69%, above the 60% minimum benchmark set for teachers to be deemed as having command of the knowledge at the grade level. The question in Figure 3.9 was classified at Grade 7 level and the reported facility was 23%.

The items in Figures 3.9 and 3.10 are also related to the structuring effect of boundary objects in a practice. The items use the nomenclature 'per cent' and 'percent'. A search of CAPS Mathematics (Senior and Intermediate Phases) documents rendered that 'per cent' and 'percent' is not used in the nomenclature with percentages in South Africa. What is used is the symbolic form, %. 'Per cent' and 'percent' are not available in the boundary objects related to calculations with percentages. Regarding the environment for percentage in school mathematics 'per cent' and 'percent' is not part of the habituated culture. Hence meaning for it had to be constructed in the situation of test-taking. It is not clear which meanings were constructed by teachers.

Discussion

I have argued here that there are some concerns about the quality of the instruments used to measure teachers' mathematics content knowledge. The outcomes of these tests are used to make policy proposals regarding directions to be taken to improve achievement in school mathematics and other mathematics tests. Spaull recommends that:

> … as a matter of urgency, they [teachers] should be required to undergo minimum-proficiency teacher training for the subjects which they teach and then re-assessed at the end of the training. Before trying to get every teacher to a desirable level, first ensure that all teachers have the basic content knowledge in the subjects that they teach. (Spaull, 2013:54)

Taylor and Taylor (2013) propagate a similar strategy of enhancing practicing teachers' mathematics content knowledge to address the conundrum of unsatisfactory performance of learners in mathematics. These suggestions have found their way into strategies adopted to address this situation. However, a meta-analysis of the impact of teacher professional development initiatives focusing attention on subject knowledge and achievement shows that the impact is minimal with an effects size of 0,09 (Hattie, 2012). This translates to the advancement of learning of 0,27 years from a learner's entry-level of knowing. This is a small return on the investments made to improve teachers' mathematical content knowledge to improve learners' achievement. More importantly, as the chapter demonstrates there are fault-lines in the tests assessing teachers' subject content knowledge. Hence, its currency for making policy recommendations, such as those made by Spaull (2013) and Taylor & Taylor (2013), should be viewed with circumspect. At a very basic level, instruments, no matter how well-conceived, cannot measure the entire universe of mathematics knowledge that teachers have and do not have. I believe that this will be hard to do. At a more substantive level, regardless of their sophistication, the procedures that are being used to reach certain conclusions on which recommendations are made depend on the quality of the instruments used to obtain the data. If these instruments do not comply with accepted criteria for 'good' tests then the conclusions are affected and lead to recommendations which, if implemented, can have unintended consequences.

Concluding remarks

A question that needs a response is whether it can be accepted that teachers have sufficient mathematics content knowledge and that the situation should remain as it is. To take such a position is foolhardy and I do not propose this view. At a broad level, mathematicians, mathematics educators, mathematics curriculum writers, mathematics curriculum material developers, mathematics teachers and generalist teachers teaching mathematics are all called upon, in various ways, to demonstrate that they have command of some piece of mathematics. But when called upon to demonstrate this knowledge it just cannot be called to mind immediately even-though there were prior education and experience related to the knowledge entity. Furthermore, it is widely agreed that all professionals do not have command of all the mathematics needed to conduct their business using mathematics. That is to say, they have mathematical knowledge gaps. For teachers, these come to fore during teaching and other school-related mathematical work in spaces such as workshops and institutes, as illustrated by Julie (2014).

The issue of teachers' school mathematics content knowledge is complex. It is not only the level of content knowledge that impacts on learners' achievement. Margaret Brown captures the complexity between mathematics content knowledge and teaching effectiveness in a memorandum to the British Parliament while parliamentarians were debating the level of content knowledge British mathematics teachers should command. She states:

> There is no simple relationship between qualifications and teaching effectiveness – some less mathematically qualified and/or less well-trained people who are intelligent, reflective and resourceful can develop considerable expertise, whereas some teachers with maths PhDs and a full PGCE fail to progress beyond adequate. (Brown, 2010)

I contend that developing teachers' mathematical content knowledge when they are engaged in activities, including teaching, directly related to their practice could effectively close whatever knowledge gaps related to what they must teach to address the unacceptable achievement of learners in mathematics. Concerning CPD initiatives, much more emphasis should be expanded on the development of teaching and it is during an engagement with such that knowledge gaps of teachers and facilitators can be more productively addressed.

References

Adam, L.A.; Oppenheimer, D.M.; Epley, N. & Eyre, R.N. 2007. Overcoming Intuition: Metacognitive difficulty activates analytic reasoning. *Experimental Psychology: General*, 136(4):569-576. https://doi.org/10.1037/0096-3445.136.4.569

Bowie, L.; Gleeson-Baird, C.; Jones, R.; Morgan, H.; Morrison, K. & Smallbones, M. 2010. *Platinum Mathematics Grade 6 Learner's Book*. Cape Town: Maskew Miller Longman.

Brown, M. 2010. *Some points relating to the teaching of mathematics*. Memorandum to British Parliament. Available: http://bit.ly/2Hykl6R (Accessed 6 June 2013).

Carnoy, M.; Chisholm, L. & Chilisa, B. 2012. *The low achievement trap: Comparing schooling in Botswana and South Africa*. Cape Town: Human Sciences Research Council Press.

DBE (Department of Basic Education). 2011. *Curriculum and assessment policy statement (CAPS): Senior phase mathematics, Grades 7-9*. Pretoria: DBE.

Essien, A. & Setati, M. 2006. Revisiting the equal sign: Some Grades 8 and 9 learners' interpretations. *African Journal for Research in Mathematics, Science and Technology Education*, 10(1):47-58. https://doi.org/10.1080/10288457.2006.10740593

Fowler, F.J. 1995. *Improving survey questions*. Thousand Oaks, CA: SAGE Publications.

Frederick, S. 2005. Cognitive reflection and decision making. *Journal of Economic Perspectives,* 19(4):25-42. https://doi.org/10.1257/089533005775196732

Freudenthal, H. 1979. Ways to report on empirical research in education. *Educational Studies in Mathematics*, 10(3):275-303. https://doi.org/10.1007/BF00314659

Giordano, F.R. & Weir, M.D. 1985. *A first course in mathematical modelling*. Monterey, CA: Brooks/Cole Publishing Company.

Hattie, J. 2012. *Visible learning for teachers: Maximising impact on learning*. London: Routlegde. https://doi.org/10.4324/9780203181522

Julie, C. 2014. Emergences and affordances as opportunities to develop teachers' mathematical content knowledge. *International Journal of Mathematical Education in Science and Technology*, 45(3):428-437. https://doi.org/10.1080/0020739X.2013.851809

McClain, K. & Cobb, P. 2004. The critical role of institutional context in teacher development. In: M. Hoines & A. Fugelstad (eds.), *Proceedings of the 28th Conference of the International Group for the Psychology of Mathematics Education* (Vol. 3) Bergen: PME. pp.281-288.

Pollak, H.0. 1969. How can we teach applications of mathematics. *Educational Studies in Mathematics*, 2:393-404. https://doi.org/10.1007/BF00303471

Reddy, V.; Visser, M.; Winnaar, L.; Arends, F.; Juan, A.; Prinsloo, C. & Isdale, K. 2016. *TIMSS 2015: Highlights of mathematics and science achievement of Grade 9 South African learners (nurturing green shoots)*. Pretoria: Human Sciences Research Council Press.

Spaull, N. 2013. *South Africa's education crisis: The quality of education in South Africa 1994-2011*. Johannesburg: Centre for Development and Enterprise.

Star, S.L. & Griesemer, J.R. 1989. Institutional ecology, 'translations' and boundary objects: Amateurs and professionals in Berkeley's Museum of Vertebrate Zoology. *Social Studies of Science*, 19:387-420. https://doi.org/10.1177/030631289019003001

Taylor, N. 2011. *The national school effectiveness study (NSES): Summary for the synthesis report.* Johannesburg: JET Education Services.

Taylor, N. & Taylor, S. 2013. Teacher knowledge and professional habitus. In: N.Taylor, S. van der Berg & T. Mabogoane (eds.), *What makes schools effective? Report of the national schools effectiveness study.* Cape Town: Pearson.

Venkat, H. & Spaull, N. 2015. *What do we know about primary teachers' mathematical content knowledge in South Africa? An analysis of SACMEQ 2007. International Journal of Education,* 41:121-130. https://doi.org/10.1016/j.ijedudev.2015.02.002

Wang, J. 2011. Re-examining test item issues in the TIMSS mathematics and science assessments. *School Science and Mathematics,* 111(7):334-344. https://doi.org/10.1111/j.1949-8594.2011.00096.x

Wu, H. 2007. '*Order of operations' and other oddities in school mathematics.* Available: http://bit.ly/32943yu (Accessed 12 January 2014).

Wynne-Wilson, W.S. 1978. Examinations and assessment. In: G.T. Wain (ed.), *Mathematical education.* London: Van Nostrand Reinhold Company. pp.181-197.

| O4 |

APPROPRIATION OF CPD BY MATHEMATICS TEACHERS: A CASE STUDY OF THE LEDIMTALI TEACHERS' APPROPRIATION OF SPIRAL REVISION

Raymond Smith, Lorna Holtman & Cyril Julie

Introduction

The design of CPD interventions must take into account how teachers learn. In any CPD intervention, it is important to promote active learning as well as constructing learning environments conducive to transform teachers' classroom practices and not simply overlay new pedagogical strategies on top of their current or traditional classroom practices. Such learning environments often involve modelling new pedagogical strategies by facilitators or colleagues. Amongst other things, constructing learning contexts for teachers to practice inside and outside the classroom allows them to reflect on the newly appropriated pedagogical knowledge and practices.

The next section discusses the notion of appropriation of new knowledge in CPD interventions.

The notion of appropriation

In the mathematics education research community, the term 'appropriation' has a deeper meaning than the dictionary definition of the word. Depending on the context the word, appropriation has different connotations, but basically, it relates to the act of taking something for oneself and using it for your own purposes. Some of the contexts in which the construct of appropriation is used include the following:

Ø   Cultural appropriation is the adoption or use of elements of one culture by members of another culture.

Ø   In law and government, appropriation is the act of setting apart something for its application to a particular usage, to the exclusion of all other uses. It typically refers to the legislative allocation of money for particular uses, such as in the context of a budget vote.

Ø   In art, appropriation is the use of pre-existing objects or images with little or no transformation applied to these objects.

In this chapter, we conceptualise appropriation as the adoption or construction of new knowledge and skills by teachers, as a way of enacting professional learning. Hence the term appropriation labels the process of constructing new knowledge and classroom practices from social and cultural sources and integrating it into pre-existing schemas to be used in the classroom.

In this sense, appropriation signifies the adoption and use of new knowledge and skills. This process may even include transforming the objects of learning to fit a particular context. Billet (1998:22) describes this aspect of appropriating as appropriating "knowledge from one situation and transforming it to have utility in another." Grossman, Smagorinsky and Valencia (1999:130) describe appropriation as: "the process through which a person adopts the conceptual and pedagogical tools available for use in particular social environments." Hence in the research on teacher professional development, the concept of appropriation may be conceived of as the process of acquiring knowledge and skills through engaging in various learning experiences, be it individually or collaboratively. It is a developmental process that comes about through socially formulated, goal-directed, and tool-mediated actions (Billet, 1998).

Expanding the notion of appropriation, Sfard (1998) proposes two complementary perspectives. These perspectives are the acquisitionist perspective (AP) and the participationist perspective (PP) of learning. Table 4.1 gives a summary of these two perspectives.

TABLE 4.1   Different perspectives on appropriation

Dimensions	The acquisitionist perspective (AP)	The participationist perspective (PP)
Goal of learning	Individual enrichment	Community building
Learning	Acquisition of something	Becoming a participant
Learner	Recipient, consumer, (re-) constructor	Peripheral participant, apprentice
Teacher	Provider, facilitator, mediator	Expert participant, preserver of discourse practice
Knowledge concept	Property, possession, commodity	Aspect of practice, discourse/activity
Knowing	Having, possessing	Belonging, participating, communication

Source Sfard (1998:6)

The AP and the PP perspectives hold the potential danger of being dichotomised. To avoid this dichotomisation, we contend that both perspectives are complementary perspectives that see learning both as agentic and socio-cultural. The appropriationist perspective is, therefore, a unifying concept for the AP and the PP perspectives. Figure 4.1 shows this relational architecture. Teacher cognition is a result of both engaging in studying or taking advanced courses as well as participating in a discourse community to advance teaching and learning in the classroom. Thus appropriation is a consequence of exercising both individual as well as relational agency by the teacher.

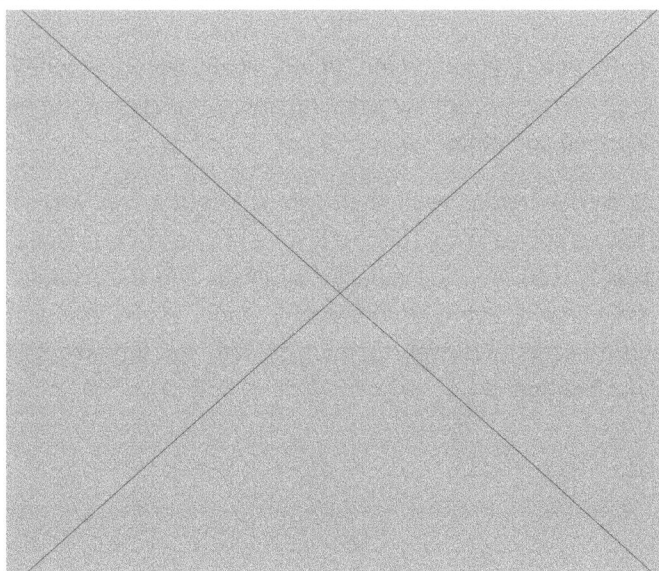

FIGURE 4.1   The architecture of appropriation

Appropriation as personal agency and social participation

Figure 4.1 illustrates the binary nature of teacher learning and brings to the fore the role of agency as well as participation in a discourse community. Merrilyn Goos describes the agentic viewpoint of teacher learning as follows: "[P]roductive tensions can emerge between teachers' thinking, action, and professional environments and that these tensions can become opportunities for teacher change." (Goos, 2013:623). In this way, she draws attention to the generative tension existing between the AP and the PP frameworks. Lerman (2002) concurs with this view and posit that sociocultural theories start from the assumption that teacher learning is about changing participation in social practices.

This dualism in the process of appropriation is further confirmed by Vygotsky's explanation that knowledge first emerges on the intersubjective plane and then subsequently operates on the intra-subjective plane. In the socio-cultural realm, Rogoff (1999) states that appropriation involves joint productive activity resulting in shared meanings for utterances and driven by a shared focus of attention. This shared meaning-making is supported by Lerman (2002) who describes professional learning as a process of initiation into a particular (scholarly) community.

Appropriation as tool-mediated learning

Säljö (2005) situates appropriation within the sociocultural perspective and explains that the learning process takes place through the mediation of knowing through tools and signs. In this chapter, we use the term 'tool' to refer to both artefacts used and produced in CPD activities as well as didactical models. Didactical models are conceived of as didactical approaches or teaching strategies which can impact the teaching and learning environment positively to improve learning outcomes.

Didactical models direct teacher-student interactions in ways that have the potential to improve learning. This improvement in learning is facilitated by acting on the didactic contract, the didactic situation and the didactic relationship. In this didactical enactment, resources, such as lesson plans and other supporting artefacts and even technology, are vital ingredients that facilitate professional learning.

Didactical models introduce certain mesogenetic[1] and topogenetic[2] techniques. Didactical models may also introduce particular chronogenetic[3] techniques (Sensevy, Schabauer-Leoni, Mercier, Ligozat & Perrot, 2005). It is these mesogenetic, topogenetic and chronogenetic features of didactical models that render them amenable to adaptation when they are appropriated by teachers and changed to fit the context.

We may, therefore, conclude that professional learning is a cognitive process which takes place through the appropriation of particular didactic models. This process is both generative and interpretive. As we will illustrate in the case of the third teacher we observed, it is this generative and interpretive practice that undergirds professional learning.

Stephen Billet supports this view and asserts that "Appropriation is the interpretative process of constructing knowledge from social and cultural sources, albeit mediated by individuals' already idiosyncratically structured knowledge" (Billet, 1998:4).

Appropriation as evidenced in the LEDIMTALI project

In the LEDIMTAL project the didactical philosophy that guides teachers' classroom practices is called the Intentional Teaching Model (ITM). According to Julie (2013), this model dictates that the goals of teaching must be clearly communicated to learners. The goals of teaching include the learning intentions as well as the success criteria. This philosophical approach is implemented through particular didactical models, such as spiral revision, assessment for learning, working with feedback and deepening mathematical thinking. For the purpose of this research, we will explore how teachers, in the LEDIMTALI project, in particular, appropriated spiral revision.

'Spiral revision' refers to an approach whereby the teacher presents short problems based on work done every day before starting to teach the lesson planned for the day. It bears some semblance to spaced or distributed practice. The difference is that spiral revision deals with small sections of the work and the intervals between revising these small sections of work are shorter.

Thus spiral revision is an approach whereby the learning process is supported by interleaved practice. This means that various topics are revised continuously. This practice is quite beneficial in the learning process. According to Sean Kang,

---

[1] Mesogenesis describes the process by which the learning environment is organised.
[2] Topogenesis describes the process of division of the activity between the teachers and the learner according to their potentialities.
[3] Chronogenetic techniques describes the teacher's way of managing the didactic time during a lesson.

"The benefit of spaced or distributed practice has been demonstrated repeatedly with a variety of study materials and learning tasks and is widely regarded as one of the most robust learning and memory phenomena in the research literature" (Kang, 2016:80). Research conducted at the University of South California showed that this approach has dramatically improved pupils' performance: "In the study reported here, the interleaving of different kinds of practice problems more than doubled subsequent test scores (77% vs. 38%) as compared to blocked practice" (Taylor & Rohrer, 2010:844).

The difference between interleaved practice and blocked practice pertains to the practice of different skills in an intermixed exercise, rather than practicing examples grouped by type. In spiral revision, this means that teachers give a short exercise or a test on previous work that does not necessarily link to the lesson to follow. For example, a set of spiral revision exercises for Grade 12 learners where the teacher is busy with teaching geometry, may look like this:

1. Solve for x and y simultaneously if

   y $= x^2 - x - 6$ and $2x - $ y $= 2$

2. Determine $f'(x)$ from first principles if $f(x) = 3x^2 - 2$
3. Prove that $\cos(90° - 2x) . \tan(180° + x) + \sin^2(360° - x) = 3 \sin^2 x$

The next set of spiral revision questions may contain single problems from yet other topics, such as finance, coordinate geometry and quadratic equations, covering the work done up to that point. In this way, spaced practice is provided that keeps the students in touch with work already completed. It also develops the skill of identifying the topic or problem type and its associated solution strategy.

To explore the process of appropriation, we turn out attention to a typology that describes certain levels of appropriation. This chapter does not use this typology as an analytic tool, or to classify the actions of the teachers involved. It is provided here merely to supplement the theoretical discussion on appropriation.

Levels of appropriation

Appropriation is described along a continuum proposed by Grossman et al. (1999). This scheme is helpful to understand and support the process of appropriation. The sub-section that follow define five levels of appropriation.

Lack of appropriation

Teachers might not appropriate a conceptual or pedagogical tool for several reasons. The concept may be too difficult to comprehend at the point in one's development when it is initially encountered. Alternatively, the concept may be too foreign to their prior frameworks at that point in their development. Teachers might also understand the concepts as intended but reject them for a variety of reasons.

Appropriation of a label

This most superficial type of appropriation occurs when a teacher adopts the name of a tool but knows none of its features. For instance, one participating teacher in our sample was familiar with the term 'spiral revision' (or quickies as it was also known amongst the teachers) and understood vaguely that it involved a speed test at the beginning of the lesson, but was not conversant with features such as interleaved practice or the conceptual underpinnings of counteracting the phenomenon of forgetting.

Appropriation of surface features

The third level of appropriation is when a teacher understands some or most of the features of a tool by their labels yet does not understand how those features contribute to a conceptual whole. For example, in the case of another teacher in the project, the spiral revision activity consisted of a three-page tutorial dealing with the same type of problems with very little variation in the structure of the problems. Learners spent an entire period working in groups. The teacher had grasped some features of revision practicum and of peer-assisted learning yet did not understand the fundamental concept of question variation and also the timespan of a spiral revision activity.

Appropriation of the conceptual underpinnings

At the conceptual level, the teacher grasps the theoretical basis that informs and motivates the use of a tool. Teachers who grasp the conceptual underpinnings of a tool are likely to be able to make use of it in new contexts and how to adapt it for personal circumstances. The cases considered in this chapter illustrate some of the principles underpinning spiral revision. This is seen in the way they organise the activity, monitor the learners' engagement and progress and provide appropriate and relevant feedback to each learner.

Achieving mastery

Mastery is the state of using the adopted tool effectively and implies the continual application of the tool in the future. This distinction contends for a continuing implementation since a deep understanding of the conceptual and pedagogical features of pedagogical tools is formed after several years of classroom practice.

In the next two sections, we consider the factors that promote the process of appropriation.

Affordances for appropriation

With respect to the process of cognition by the professional person, Grossman et al. (1999) suggest that appropriation follows three phases:

1. Growing into submission,

2. submitting, and

3. growing out of submission.

For teacher professional learning, these phases may be facilitated by recognising that teachers will:

Ø   appropriate from a reputable resource;

Ø   engage in a process of meaning-making; and

Ø   adapt the new knowledge for use in their specific context.

These three affordances highlight some critical factors in teacher professional learning. In the first instance, it highlights the fact that teachers can recognise quality in facilitation and content. Secondly, they are aware of the importance of participation and collaboration as the pre-requisites of professional learning. Thirdly, it emphasises that professional learning is always purposeful and purpose-driven. In other words, engaging in professional learning is not for the sake of compliance with bureaucratic prescripts, but an inherent component of professionalism.

Factors affecting the degree of appropriation

Having noted the important role that agency and a professional disposition play regarding the appropriation of professional learning, we also briefly indicate other salient points in this regard. There are a variety of factors that can affect the level of appropriation by the teacher (Grossman et al., 1999). Some of these factors are agency, context, individual attributes, motivation, the interest and knowledge of the teacher as well as resource provisioning.

According to Grossman et al. (1999), appropriation is enhanced when participants play an active role in their learning process. In this regard, a more positive learning environment will most likely result in higher degrees of appropriation. This is positively influenced by the beliefs, goals, expectations, and motives of the teacher. They further contend that teachers who are intrinsically motivated are more likely to reach a higher level of appropriation. However, the teaching biography of the teachers as well as their knowledge of the subject and learning theories may play an enhancing role in the process of professional learning. In this respect, theories refer to both local and global theories of learning, whether explicitly or implicitly held.

As has been indicated, appropriation is tool-mediated (Billet, 1998). As humans, we rely on material artefacts as repositories of and resources for thinking and learning. Access to a broad range of resources for improving teaching and learning, including ICTs and the types of boundary objects discussed in Chapter 7, play a significant role in the process of professional learning.

Instances of appropriation of spiral revision as evidenced in the LEDIMTALI project

The data used in analyses of the case studies that follow were obtained through classroom observation of teachers in action using a video recording for analysis (Schoenfeld, 2013, 2016, 2017). The analysis was situated in a qualitative research tradition and followed an ethno-methodological approach.

Case 1: Lillian

Lilian[4] teaches a Grade 10 class in a school located in a low socio-economic status (SES) community. She has 38 learners in her mathematics class and the language of instruction is English. This is different from the learners' home language.

She uses a 'quickie' to assess the learners' knowledge of the sides of the triangles (in numerical format) normally used to represent the special angles in trigonometry. As previously stated, the term 'quickie' was coined in the project and denotes a short spiral revision activity that precedes a lesson. The activity does not have to relate to the subsequent lesson.

The use of quickies is a technique a teacher uses to ensure that previously taught concepts and procedures are consolidated and retained in the learners' long-term memory. It also supports the development of a triad of metacognitive

4   Pseudonyms are used in all three case studies.

strategies namely the learner's strategy knowledge, the learner's task knowledge and the learner's personal knowledge (cf. Garfalo and Lester, 1985; Gurat and Cesar, 2016; Pintrich, 2002). Strategy knowledge refers to the learner's knowledge about strategies, including general and specific cognitive strategies, as well as knowing how and when to apply them. This is further supplemented by the learner's task knowledge base developed through experiences of having done certain types of problems, and the learner's levels of self- knowledge and -confidence are bolstered.

The spiral revision session proceeded as follows:

Teacher   As part of revision I want you to complete the following table in five minutes.

The teacher then drew the following table on the writing board.

Complete Trig Ratios

	y	r	x
0°			
30°			
45°			
60°			
90°			

FIGURE 4.2   Spiral revision exercise written on the board

After some time the teacher walked among the learners to observe their answers. Figure 4.3 is an example of how she gave feedback. The teacher shows the thumbs-up sign to communicate to learners that their work is correct as shown.

FIGURE 4.3   The teacher used the thumbs up sign to show approval of learners' work

She sampled one row by looking at the work of the learners on her right-hand side. After she surveyed the learners' work in this row, she wanted to ensure that the rest of the class confirmed that their work was correct. She requested the learner to come to the board and complete the table of values, by stating:

Teacher   I want a volunteer to come and complete the table on the whiteboard.

Figure 4.4 shows the volunteer learner's completed task. The teacher then asked the entire class to mark their own work accordingly. This allowed learners to correct their work.

Complete Trig Ratios

	y	r	x
0°	0	1	1
30°	1	2	$\sqrt{3}$
45°	1	$\sqrt{2}$	1
60°	$\sqrt{3}$	2	1
90°	1	1	0

FIGURE 4.4   Solution to spiral revision exercise written on the board by a learner

The entire activity lasted three minutes, which is within the allotted time for a 'spiral revision' activity. She then proceeded with the lesson planned for the day. The teacher thus ensured that the she reminded learners about the angles and sides used to determine the trigonometric ratios for the special angles, namely the trigonometric ratios of 0°, 30°, 60° and 90°. This regular and consistent repetition of important theory helps the learners to commit facts and concepts to their long-term memory.

Case 2: Bonita

Bonita teaches a Grade 12 class in a school located in a low-SES community. Her mathematics class is fairly small for her context with no more than 15 learners. Her 'quickie' is a form of a 'spiral revision' activity focused on mastery of the skills and knowledge involved in solving different types of quadratic equations. This is obvious from Figure 4.5, which shows the problems assigned to the class.

The first type of problem is focused on the 'zero-product' principle. The second problem is intended to remind and assess the learners' strategic knowledge of solving an equation where a surd form appears. The solution strategy for this type of equation is very different from the normal way in which the solution of a quadratic trinomial equation is approached. This variation of problem structure is supported in the literature (cf. Kang, 2016).

Figure 4.5 shows how the task was projected onto a screen using a data projector linked to Bonita's laptop computer.

Solve for x

$(7x - 5)(x + 5) = 0$

$\sqrt{x - 2} + 2 = 3$

FIGURE 4.5   Spiral revision exercise

The revision session lasted five minutes and the teacher observed the learners as they were engaged in solving the two equations on the improvised 'whiteboards'. The 'whiteboard' consisted of a laminated A4 sheet of paper on which the learners could write and erase their work. They were also useful for the learners to show their answers, as is illustrated in Figure 4.6.

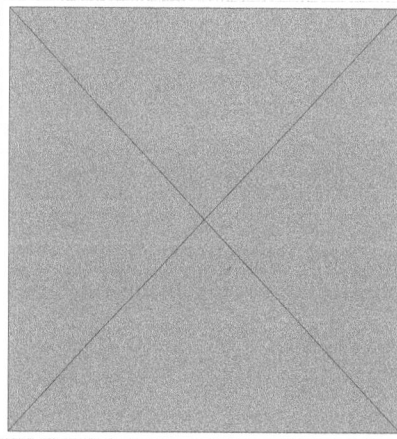

FIGURE 4.6   Learners showing their answers using the improvised whiteboards

After observing the learners' responses the teacher voiced the answers and then directed her attention to one specific learner who probably did not know how to do the second problem. She said to this learner: '*Carefuuuul*' in a high-pitched intonation and extended the last syllable as a way to emphasise her concern.

When the teacher walked back to her desk, this learner leaned over to a classmate and looked at the correct answer. In other words, the learner, in this case, used her peer to compare and correct her work.

The teacher concluded this five-minute session by declaring: "*Excellent work, except for two of you who breaks my heart*". There were congratulatory support and confirmation, as well as empathy in her declaration. This is a sign that she cares about the progress of each learner. In addition, the two learners in whom she expressed disappointment know that they need help and will approach either the teacher or one of their peers to fill in the gap in their understanding. It is obvious from the video analysis that they indeed sought the assistance of one of their peers.

Case 3: Donald

Donald teaches a Grade 10 class with 45 learners in a fairly big school located in a low-SES community. The learners are taught in a language other than their mother tongue. In his case, the data emanates from a colloquium presentation reporting on his efforts and experiences in implementing the didactical strategy of spiral revision. After three years of involvement in the project, a two-day colloquium was held. At this colloquium, some teachers reported back on their experiences related to the implementation of particular strategies. Donald reported back on his experiences regarding the implementation of spiral revision. This report serves as an instantiation that appropriation of this didactical model has occurred.

This case is important for the discourse about appropriation because it illustrates important features of appropriation. One noteworthy aspect of Donald's practice is that it provides evidence of how a theoretical construct, such as spiral revision, is transformed when it 'travels' from the site of teacher development to the site of teacher practice. Daley (2001) investigated this phenomenon and concluded that appropriating new knowledge is a recursive, transforming process rather than directly mimicking the new knowledge in a different context. Donald's appropriation of this pedagogical strategy illustrates this transformative aspect of appropriation. The series of slides that come from his presentation at a colloquium narrates his experiences.

The series of slides shown in Figure 4.7 summarise his narration at the colloquium. The slides are provided to allow Donald to speak for himself.

What did I learn in the project?	**How does it work in my class?**
ø Implementing spiral revision.	ø I use the first 10 minutes of my class.
ø Spiral revision is a strategy to make sure that learners do not forget previously taught work.	ø I use the cards that I got from the project for Grade 9.
ø The means that with every lesson I include a problem from previous lesson.	ø I did this for two weeks, but I then adapted the programme because some learners could not keep up with the pace.
	ø I now have a dedicated period on a Friday for spiral revision.
1	**2**
Adaptation	**Outcomes**
ø I felt it was necessary to adapt the cards.	ø This strategy resulted improved performance in the ANAs.
ø I made my own cards because I wanted to differentiate.	ø Grade 10 and 11 results also improved dramatically.
ø This also prevented learners from copying from each other.	ø More learners showed interest in mathematics.
3	**4**

FIGURE 4.7   Presentation slides at the colloquium: The teacher's narration

The first three slides tell the story of adopting and adapting the pedagogical model. They show how he overcame the challenges presented by his local school context. It is noteworthy that the adaptation he made was as a consequence of being aware that learning is important. As a result, the way he adapted the strategy was geared to maximising the learning opportunities for his learners.

The interpretive and generative aspects of appropriation are evident from the next series of slides shown in Figures 4.8 and 4.9.

Grade 9 **Revision Exercises** **Card 4**

Real Numbers; Exponents1

1. Is $(2x)^2 + 3x^2 = 7x^4$ correct? Explain your answer.

2. Arrange the following in descending order:

   $4, 03; 4,33; 4\frac{1}{3}; 4,3$

3. Simplify: $\dfrac{3a^{-2}b \times 24b^{-1}a^{-1}}{9a^{-4}b^{-3}}$

FIGURE **4.8** Example of a project-designed spiral revision card

Factorise fully

1. $5p^3 - 15p^2$

2. $4 - y^2$

Simplify

1. $(2x - 3)(x+4)$

2. $\dfrac{2x^2y^6}{(2x^3y^2)^3}$

1. Consider the sequence: 17; 13; 9; ...
   1.1 Write down the next 3 terms.
   1.2 Determine the general formula, T_n.
   1.3 Determine the 22nd term of the sequence.
   1.4 Which term is equal to -19?

2. Temoso invested R 1 500.00 for two years at a rate of 11% simple interest (S.I) per year. What is her investement worth at the end of the second year?

3. Solve for x: $x - \dfrac{x-1}{2} = 3$

1

13

1. Factorise fully: $2x^2y^2 - 4x^2y + 10xy^2$

2. Factorise fully: $2^3 - 8x$

3. Calculate the simple interest on R5 400 at 6% per annum for 4 years.

4.1 Write down the next TWO terms in the sequence: 3; 8; 13; ...

4.2 Write the general term of the sequence T_n.

4.3 Which term in the sequence is equal to 38.

18

1. ABCD is parallelogram Calculate the value of x and size of $\hat{B}$.

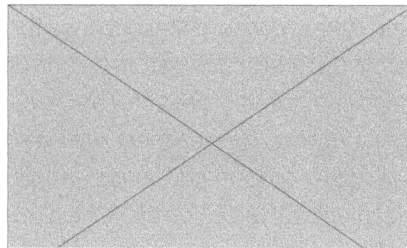

2. Simplify: $\dfrac{3a^{-2}b \times 24b^{-1}a^{-1}}{9a^{-4}b^{-3}}$

3. Factorise: $81 - 100y^2$

22

FIGURE **4.9** Spiral revision cards developed by Donald

Discussion

These three teachers appropriated spiral revision in varying ways. They observed how it was modelled during the teacher institutes and was willing to experiment in their classrooms. It is also clear that they understood the pedagogical benefits of this didactical model and is committed to experimenting with this strategy.

The first observation is that they used their initiative and extended the resources provided in the project. They did not use the cards they received from the project. In other words, the teachers did not rely solely on the revision cards provided by the project. This is an important observation regarding the consequences of professional learning through an appropriation of the tools, techniques and strategies presented in the project. When this takes place, professional learning manifests and we can claim that the project has made a lasting impact. The process of appropriation has helped teachers to grow professionally and extend their compendium of instructional strategies.

The Grade 10 teacher (Lillian) probably copied a table which occurs in the textbook making use of the primary resources provided by the education department. This is important because it shows how the teacher can integrate the various resources at her disposal in the course of implementing newly appropriated didactical tools.

The Grade 12 teacher (Bonita) used a short revision exercise sourced from the internet and projected it onto a screen using a laptop and data projector. This way of enacting the implementation of a particular didactical model shows that they do not rely on the project to supply materials, but they can employ the resources at hand. This is encouraging because the teachers did not attempt to mimic what the project leader demonstrated, nor did they solely depend on the resources provided through the project. This is an important instantiation of the process of appropriation. The teachers also demonstrated an awareness of contextual factors and constraints when implementing strategies adopted through engagement with a CPD project. Hence, they considered what is possible in their respective contexts and how they could adapt to the situation.

Merriam and Leahy (2005) describe the work environment as a learning environment that also impacts on the way teachers appropriate new knowledge and skills. This is because teachers need to plan, and in this planning process, anticipate possible implementation challenges. They furthermore conclude that the more authentic the professional learning experience, in other words, the closer it is aligned to actual practice, the more likely that it will be appropriated by teachers.

We may summarise the process of appropriation by referring to the way we see it unfolding in the LEDIMTALI project, as the pedagogy of enactment and inquiry. The pedagogy of enactment contributes to a learning cycle based on reflection. This initiates a pedagogy of inquiry, leading to the adaptation of the new knowledge to become situated and relevant. In this way, we may ensure that appropriation of new knowledge surpasses the appropriation of surface features and grows to the level of mastery based on a deep understanding of the conceptual underpinnings of the new practices.

Conclusion

Our observations of teacher professional learning through the process of appropriation evidenced the possibilities for CPD. The teachers observed in the three case studies continually moved between cognition and external symbolic tools in their meaning-making endeavours. This happened as they recalled the demonstration lessons by the field workers in the project and depended on the accompanying artefacts or the classroom resources provided by the project as prototypes. A pedagogical consequence of this is that artefacts and resources used or produced during CPD activities and subsequently at school represent important support during the process of appropriation. Knowing when and how to rely on artefacts, whether provided by the CPD project, self-made or borrowed from others, is an important kind of meta-knowledge in the process of appropriating new teaching strategies.

References

Billet, S. 1998. Appropriation and ontogeny: identifying compatibility between cognitive and sociocultural contributions to adult learning and development. *International Journal of Lifelong Education*, 17(1):21-34. https://doi.org/10.1080/0260137980170103

Daley, B.J. 2001. Learning and professional practice: A study of four professions. *Adult Education Quarterly*, 52:39-54. https://doi.org/10.1177/07417130122087386

Garofalo, J. & Lester, F.K. 1985. Metacognition, cognitive monitoring, and mathematical performance. *Journal for research in mathematics education*, 16(3):163-176. https://doi.org/10.2307/748391

Goos, M. 2013. Sociocultural perspectives in research on and with mathematics teachers: A zone theory approach. *ZDM – The International Journal on Mathematics Education*, 45(4):521-533. https://doi.org/10.1007/s11858-012-0477-z

Grossman, P.L.; Smagorinsky, P. & Valencia, S. 1999. Appropriating conceptual and pedagogical tools for teaching English: A conceptual framework for studying professional development. *Report Series*, 12011:1-27.

Gurat, M.G. &, Cesar, T. 2016. Metacognitive Strategy Knowledge Use through Mathematical Problem Solving amongst Pre-service Teachers. *American Journal of Educational Research*, 4(2):170-189.

Julie, C. 2013. Towards a model for intentional teaching for improving achievement in high-stakes mathematics examinations. In: Z. Davis & S. Jaffer (eds.), *Proceedings of the 19th Annual Congress of the Association for Mathematics Education of South Africa*, Vol. 1. Cape Town: AMESA. pp.86-96.

Kang, S.H.K. 2016. The benefits of interleaved practice for learning. In: J.C. Horvath, J. Lodge, & J.A.C. Hattie (ed.), *From the laboratory to the classroom: Translating the learning sciences for teachers*. New York: Routledge. pp.91-105.

Lerman, S. 2002. Cultural, discursive psychology: A sociocultural approach to studying the teaching and learning of mathematics. In: C. Kieran, E.A. Forman & A. Sfard, *Learning discourse: Discursive approaches to research in mathematics education*. Dordrecht, The Netherlands: Kluwer Academic Publishers. pp.87-113. https://doi.org/10.1007/0-306-48085-9_3

Merriam, S.B. & Leahy, B. 2005. Learning transfer: A review of the research in adult education and training. *PAACE Journal of lifelong learning*, 14:1-4.

Pintrich, P.R. 2002. The role of metacognitive knowledge in learning, teaching, and assessing. *Theory into practice*, 41(4):219-225. https://doi.org/10.1207/s15430421tip4104_3

Rogoff, B. 1999. Cognitive development through social interaction: Vygotsky and Piaget. *Learners, learning and assessment*. pp.69-82.

Säljö, R. 2005. *Learning & cultural tools: On processes of learning and the collective memory*. Stockholm: Norstedts Akademiska Förlag.

Schoenfeld, A. 2013. Classroom observations in theory and practice. *ZDM – The International Journal on Mathematics Education*, 45(4):607-621. https://doi.org/10.1007/s11858-012-0483-1

Schoenfeld, A. 2016. Making sense of teaching. *ZDM – The International Journal on Mathematics Education*, 48(1):239-246. https://doi.org/10.1007/s11858-016-0762-3

Schoenfeld, A. 2017. Uses of Video in understanding and improving mathematical thinking and teaching. *Journal of Mathematics Teacher Education*, 20(5):415-432. https://doi.org/10.1007/s10857-017-9381-3

Sensevy, G.; Schubauer-Leoni, M.L.; Mercier, A.; Ligozat, F. & Perrot, G. 2005. An attempt to model the teacher's action in the mathematics class. In: *Beyond the apparent banality of the mathematics classroom*. Springer, Boston, MA. pp.153-181. https://doi.org/10.1007/0-387-30451-7_6

Sfard, A. 1998. On two metaphors for learning and the dangers of choosing just one. *Educational Researcher*, 27:4-13. https://doi.org/10.3102/0013189X027002004

Taylor, K. & Rohrer, D. 2010. The effects of interleaved practice. *Applied Cognitive Psychology*, 24(6):837-848. https://doi.org/10.1002/acp.1598

| 0 5 |

FACILITATING AND MEDIATING BY MATHEMATICS TEACHER EDUCATORS AS A MATTER OF POSITIONALITY

Faaiz Gierdien

Introduction

LEDIMTALI involves participants who have different perspectives and hence positions with regards to the teaching and learning of school mathematics. This is due to them coming from different institutions, such as higher education institutions, the provincial Education Departments, and schools locatedtiat in areas that are historically and socio-economically challenged. These participants include mathematicians, mathematics teacher educators, mathematics teachers (hereafter 'teachers') who teach in Grades 10-12, namely the FET band, curriculum specialists and mathematics curriculum advisors (see Figure 5.1). The different levels of expertise among the participants imply that they think differently about the teaching and learning of school mathematics. A major task of mathematics teacher educators is finding ways to facilitate and mediate developments, ideas and goals related to the LEDIMTALI project, to the teachers. This chapter, therefore, addresses the question: What is there to learn from the facilitating and mediating that mathematics teacher educator's do in relation to the teachers with whom they work?

For any facilitating and mediating to happen, mathematics teacher educators have to be aware of the teachers' classroom realities. These have to be mapped with respect to the education context, especially regarding mathematics teaching in the last three years of non-compulsory schooling. The teachers work in high schools located in segregated neighbourhoods that were historically enforced under apartheid. In the public eye, the teachers are negatively positioned and have to deal with pervasive deficit discourses (Keitel, 2005). They have to contend with the performance of their schools in the high-stakes NSC Mathematics examinations. The teachers are 'targeted' (Labaree, 2011) through remarks that they are not as dedicated as the teachers in the more affluent, historically advantaged schools (Spaull, 2013). The latter schools, mainly located in the leafy suburbs of major South African cities, were reserved under apartheid laws for white learners.

Given the context of these high-stakes examinations, it is not surprising that the teachers take the NSC Mathematics examinations, and to a lesser degree, the ANAs, as their major 'reference point' (Boardman & Woodruff, 2004). In other words, the teachers are "put to the test" with respect to their learners' performance (Smith, 1991:8). A particular result is a tension between the reference points of teachers and mathematics teacher educators.

Central to the problem statement for this chapter is that in CPD tasks, there may arise 'tension' between mathematics teacher educators (or facilitators) and teachers. They represent two distinct but overlapping modes of practices or ways of working. Teachers may choose to focus only on the particular task at hand, whereas, mathematics teacher educators may see the task as presenting a 'ripe moment' for introducing an extension of the task. The different positions outlined point to the importance of the mediating and facilitating role of the mathematics teacher educators. Figure 5.1 represents the participants and the ways they relate to each other.

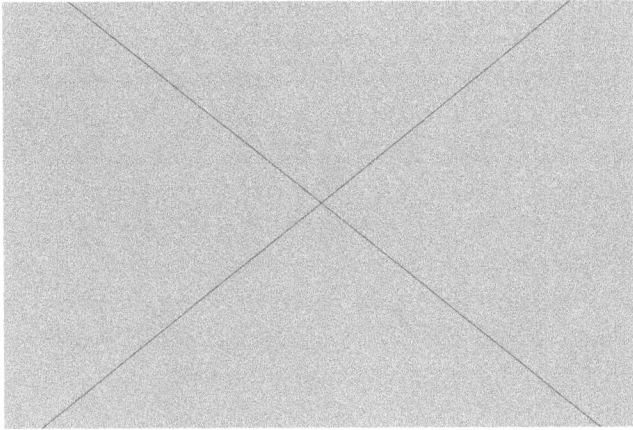

FIGURE 5.1 Outline of the differentially distributed expertise among participants in LEDIMTALI

The arrows, going both ways, signify possible contact, conversations and relationships between the participants. For example, there can be contact between the teachers and mathematicians, as well as between the teachers and the mathematics teacher educators and curriculum specialists. The participants work at institutions that are physically separated, such as higher education institutions, schools and the provincial Department of Education offices. Also, the participants have different and at times overlapping ways of knowing, talking and doing school mathematics. The various arrows in Figure 5.1 point to participants' distinct professional practices that come together through activities organised by the LEDIMTALI project. Given this problem statement and situation, it becomes important to understand what mediating and facilitating on the part of mathematics teacher educators in relation to teachers, entail.

Facilitating and mediating

Facilitating and mediating are interrelated actions and interactions, meaning that one is connected to the other (Kennedy, 2016:955; Poekert, 2011). In LEDIMTALI project activities, facilitating and mediating are instances that involve organising and affording the teachers and the mathematics teacher educators opportunities to learn. These could, for example, include doing and designing examination practice questions and different teaching units. In particular, facilitating is the physical presentation or provision, to the teachers, of copies of sample questions, blank sheets of paper, and guidelines for developing teaching units. Facilitating also concerns the provision of whatever computer software or paper is needed to record notes and diagrams that emerge during discussions between teachers and mathematics teacher educators who are in a particular group.

Mediation refers to interactions, such as discussions and clarification of issues directly or indirectly related to the mathematics content, between teachers and mathematics teacher educators. The issues discussed could focus on the rationale for approaching, doing and extending particular questions related to the mathematics content. It may then be the case that the mathematics teacher educators and teachers hold similar or different views. Mediating of the mathematics content can be facilitated employing pencil and paper technology, which, for example, could be used to draw diagrams or 'sketch graphs' of quadratic functions. Pencil and paper technology can be compared and contrasted with other technologies, such as the chalkboard on which diagrams or quadratic functions can be drawn (see Figure 5.3).

During any facilitating and mediating the mathematics teacher educator takes on various roles of positionality, one of which is that of researcher positionality (Milner, 2007). This means that she or he has to look out for learning opportunities to initiate and predict discussions (Wall, 2000). Invariably this is done from the position as a researcher. In these activities, and learning on the part of mathematics teacher educators and teachers cannot be assumed to happen automatically and it has to be set up.

Facilitating, mediating and 'boundary objects'

There is analytic value in the notion of a 'boundary object' when it comes to mediating and facilitating in any work arrangements between teachers and mathematics teacher educators (Marick, 2001; Star, 1988; Star & Griesemer, 1989; Wenger, 2000:236). In a metaphorical sense, a 'boundary' refers to a periphery or an edge. It is thus not difficult to see that in the activities of the LEDIMTALI project there is a metaphorical boundary between the distinct but overlapping modes of practice of the teachers and mathematics teacher educators. In ordinary language, an object is a 'thing' or material entity that consists of more or less well-structured stuff (Star, 2010:603). An object can also be conceptualised as 'reification' (Wenger, 1998) in the form of a concrete object that embodies a set of ideas or processes (Stein & Coburn, 2007:5). In the case of the LEDIMTALI project, there is the example of 'concrete', (printed) teaching units.

This notion of boundary object has gained prominence in the professional development literature (Cobb, McClain, Lamberg & Dean, 2003; Cobb, Zhao & Dean, 2009; Stein & Coburn, 2007). This potential stems from the fact that deploying boundary objects can facilitate the possibility of shared meaning-making (Star & Griesemer, 1989). This does not necessarily imply that there will be consensus with respect to the boundary objects, between teachers and mathematics teacher educators. Wenger (2000) presents different 'categories' of boundary objects, while Star and Griesemer (1989) differentiate 'types' of boundary objects. In terms of this chapter, the applicable category is 'discourses', and in particular, the language associated with quadratic sequences and quadratic functions. The applicable type is 'repositories' of which the high-stakes NSC past examinations papers and mathematics textbooks are examples because they 'store' knowledge (Star, 1988:48).

Why refer to teachers' and mathematics teacher educators' positionalities?

In any facilitation and mediation in the CPD activities, the teachers and the mathematics teacher educators take on particular positionalities, which, in part, is adapted from Milner's (2007) notion of researcher positionality. For example, mathematics teacher educators as researchers have to be aware of their positionality in relation to the challenging socio-economic contexts in which they work. Milner avers that there are dangers – seen, unseen and unforeseen – in mathematics teacher educators claiming that they know the mathematics content in relation to the teachers and that they know what it is like to work under those contexts. They are not employed at and do not directly report to the schools involved in the CPD LEDIMTALI project in ways that the teachers do. The teachers, as described earlier (see Spaull, 2013), on the other hand, have to contend with the associated deficit discourses. Also, depending on the learning opportunities available, the teachers will need time to make connections between different forms of representations of mathematical concepts, such as a quadratic sequence and the quadratic function. Mathematics teacher educators have vantage points with respect to quadratic and other functions because of the kinds of literature that they are required to be familiar with as part of their job descriptions in the higher education institutions where they work.

This content example is representative of mathematics education researcher (MER) mathematics (Sfard, 1998). The teachers, on the other hand, take as their major reference point or positionality, the legitimate school mathematics, namely the mathematics associated with high-stakes, time-restricted examinations, which is for them an important boundary object. The use of 'perspectives' is not considered as suitable because teachers may agree with the mathematics teacher educators' perspectives but might position themselves so that they can deal with ways the legitimate school mathematics is structured, which is a key feature of the schooling system and its attendant pressures.

Towards an analysis framework

When analysing any facilitating and mediating 'moments' in the activities of the LEDIMTALI project, the notions of boundary objects and mathematics teacher educators' positionings have to be considered. During these 'moments' at least two 'kinds of mathematics' come into the picture. To develop a framework, these notions will have to be reviewed in an integrated way.

The activities of the LEDIMTALI project have to contend with mathematics represented in boundary objects, such as high-stakes assessments, school textbooks and in the journals that mathematics teacher educators read. School mathematics is a particular or special kind of mathematics (Julie, 2002; Watson, 2008) that has been institutionalised in the school and is subject to its institutional and epistemological contexts, namely ways of knowing and doing peculiar to schools. In the context of the high-stakes examinations described earlier, the epistemology with respect to school mathematics content is legitimated by tests and examinations. Thus the content of the NSC Mathematics examinations becomes 'high-stakes'. Drawing on Kvale's (1993:222) analysis of the knowledge associated with examinations, Julie (2012:1) makes the case for the notion of "legitimate school mathematics knowledge", which is the knowledge associated with high-stakes, time-restricted examinations. On the other hand, mathematics teacher educators bring a particular perspective to school mathematics that can be characterised as MER mathematics (Sfard, 1998). This mathematics draws on insights from learning theories, such as constructivism or socio-constructivism, and pedagogy, namely from different strands of literature emanating from and related to the domain of mathematics education. One definite result is that mathematics teacher educators face dilemmas when working with teachers. Mathematics teacher educators may want to emphasise the importance of representing the function concept graphically, algebraically and through the use of a table as a 'unifying concept'. This same (linear or quadratic) function concept can be conceptualised and represented over the set of integers or the set of real numbers. There are numerous studies on the centrality of the function concept in school mathematics (see Freudenthal, 1983; Nyikahadzoyi, 2006). To date, we know about teachers from similar education contexts who work with the 'function concept' (Gierdien, 2014). A focus on this as a 'unifying concept' may not go well in school mathematics, because it would require vertical coordination across the grade levels (Watson, 2008:6). When facilitating and meditating during teacher institutes, mathematics teacher educators will thus need to look for opportunities where there can be overlaps with their MER mathematics and ways the teachers can come to think about an intentional teaching model.

When doing research in mediating and facilitating, the mathematics teacher educator takes on a particular 'researcher positionality' (Milner, 2007:388). The mathematics teacher educator cannot view the teachers' 'reference point', described earlier, as categorically negative. The teachers in the LEDIMTALI project and like those from affluent schools are influenced by a didactical transposition of the mathematics and its high-stakes consequences for examinees (Brousseau, 1997:21; Harley & Wedekind, 2004; Teese, 2000). Ideally, the mathematics teacher educator's positioning is one of searching for and capitalising on opportunities, together with the teachers, to see where and how legitimate school mathematics knowledge can be expanded and connected in ways that show interrelationships between concepts. The mathematics teacher educator's vantage point positioning is about learning what it takes to forge an agenda for 'deep progress' (Watson & De Geest, 2005; Watson, De Geest & Prestage, 2003;) and 'spiral revision' (Selter, 1996), where there is "recurrent practice of work covered" (Julie, 2013) before or a 'productive practice' and ways of explicating intentional teaching for school mathematics. Deep progress and spiral revision can mean drawing up and extending tasks that point out mathematical connections between quadratic sequences as algebraic expressions peculiar to sequences and series and parabolas as quadratic functions.

To analyse facilitating and mediating empirically in terms of data, the analytic framework shown in Figure 5.2 will be justified and used. This framework has to include the two positionings mentioned above. It also has to indicate the CPD context, whether it be a teacher institute, workshop, examination practice or the high-stakes education context described earlier.

Facilitating and mediating in a high-stakes examinations context of CPD

Positionality: legitimate school mathematics teachers

Positionality: MER-mathematics, vantage point of mathematics teacher educators

Mathematics content: Quadratic functions

Boundary objects: teaching units and examination practice questions taken from past papers

FIGURE 5.2   Analytic framework for facilitating and mediating in the LEDIMTALI CPD model

The left-hand column shows issues that the mathematics teachers are concerned with, namely their reference point(s). The right-hand column indicates dilemmas and issues that the mathematics teacher educator brings to the workshops and teacher institutes.

Data and analysis

Data collection and analysis used to answer the research question are qualitative and informed by the analysis framework (See Figure 5.2). Data analysis is based on comparing the teachers' positionings with respect to the mathematics content that appears in the activities of the LEDIMTALI project to my positioning as the mathematics teacher educator. This is influenced mainly by MER-mathematics and boundary objects.

Central to doing any analysis of data is that during teacher institutes or other types of contact with the teachers, the legitimate school mathematics content comes into contact with MER-mathematics. An important result is that there are two positionings with a metaphorical boundary between the two. There is also the presence of boundary objects. Because of this, data analysis will be done using a 'constant comparison' between the two different positionings – teachers' and mine with respect to functions (Glaser & Strauss, 1967:102). There is also the boundary between the two worksites, namely the teacher institutes that which are held off-campus and the teachers' classroom that will have to feature in the analysis.

The data incident that will be considered has two parts. The first concerns a question about finding the formula for a quadratic sequence, as an algebraic expression, which happened during a teacher institute on examinations practice at an off-campus venue. Here I was assigned to work with a group of teachers. The second part shows two 'still shots' from a videotaped revision lesson from a teacher's classroom or worksite. They show a learner's solution procedures, guided by the teacher, for solving a quadratic inequality. The analysis focuses on different positionings, the teachers' and mine, with respect to the symbolic, graphical and tabular representations of the quadratic function concept.

Positionalities with respect to quadratic functions

For purposes of highlighting facilitating and mediating, this section is organised as follows. Firstly, it includes a description of the data incident as a way of showing where and how the two positionings come into the picture. This is followed by an analysis that takes into account the worksite of the teacher institute in relation to the teachers' main worksite (their classroom) and the gaps in the sequencing of questions on quadratic functions as they appear in the boundary object (the high-stakes NSC Mathematics examinations questions). It shows that the teachers are familiar with multiple (symbolic and graphical) representations of quadratic inequalities and functions with their turning points over the domain of real ($R[x]$) numbers. As a boundary object, these examinations, in particular, shift a tabular representation of quadratic functions out of purview in the case of quadratic sequences and series. In this case, the quadratic expression for the nth term is viewed as an algebraic expression over the domain of integer ($Z[x]$) numbers.

During one of the teacher institutes on an NSC Mathematics examinations practice module, I was assigned to work with a group of teachers where we discussed approaches to doing, recognising and drawing up questions and answers on quadratic sequences. This module development is key in terms of getting the teachers familiar with the pervasive high-stakes, time-restricted NSC Mathematics examinations. This entailed legitimate school mathematics knowledge as well as the mathematics teacher educator's perspective on this knowledge. From my vantage point as a mathematics teacher educator, I was and am aware of structural connections within and between quadratic sequences and the solving of quadratic inequalities in terms of symbolic, graphical and tabular representations of quadratic functions. I had an opportunity to mediate the teachers' main reference point – the high-stakes NSC Mathematics examinations, a boundary object – with respect to quadratic functions.

More details on this opportunity or 'moment' is provided to show how teachers understand structural connections with respect to quadratic sequences and functions in terms of the domain of integer ($Z[x]$) numbers. As a group, our task was to write different types of questions and answers for this module. The first question we addressed was to identify whether a sequence is quadratic or not. In one example, we dealt with the following sequence:

…-4, 4, 10, 14, 16…

This sequence can be represented through a table shown below.

TABLE 5.1 A quadratic sequence showing its first and second differences

Term number (T_n)	1	2	3	4	5
Sequence	-4	4	10	14	16
1st difference		8	6	4	2
2nd difference			-2	-2	-2

This table is not exactly what we as a group drew up but rather my reconstruction of how we agreed that there was a constant second difference. We concluded that the sequence is quadratic because of the constancy of the second difference.

Through our calculations, we found the expression for the quadratic sequence. To be in line with mediating, I showed that in my extension (see Table 5.2) of this table there is a 'turning point' in the sequence. I then pointed out that the quadratic expression we found can also be viewed as the 'equation' for a parabolic or quadratic function in the domain of real ($R[x]$) numbers. In the words of Freudenthal (1983:492) we can speak of 'the equation of a function' – in this case, a quadratic function.

I also showed the group a table similar to the one reconstructed in Table 5.2 and an accompanying hand-drawn 'sketch graph' of the related 'parabola'. Using the available paper, I made a graphical, tabular and symbolic representation of the quadratic sequence. I then pointed out where the turning point of this quadratic function lies and what I saw as mathematical connections through multiple representations, between quadratic sequences and the graphical and tabular representation of the related parabola. Table 5.2 is an extension of Table 5.1 and shows how this quadratic sequence, consisting of italicised integer values, was extended.

TABLE 5.2 Mathematics teacher educator input (italicised) as an extension of the quadratic sequence in Table 5.1

					'Turning point' somewhere between term number 5 and 6					
Term number (Tn)	1	2	3	4	5	6	7	8	9	10
Sequence	-4	4	10	14	16	16	14	10	4	-4
1st difference			8	6	4	2	0	-2	-4	-6	-8
2nd difference				-2	-2	-2	-2	-2	-2	-2	-2

The teachers noted but were not willing to consider my reference to the presence of a turning point and the associated parabola or a quadratic function in the domain of real ($R[x]$) numbers.

If we look at the teachers' main worksite – the classroom – we see that they are familiar with the discourse or language, namely symbolic and graphical representations of quadratic functions and their turning points, evident in learners' work, in ways that are separated from quadratic sequences. This familiarity is determined primarily by what needs to be taught as the academic year progresses and by the sequencing of questions found in the boundary object, the high-stakes NSC Mathematics examinations questions. To illustrate this mathematical connection between quadratic algebraic expressions and quadratic functions, we analyse two 'snapshots' taken from a revision lesson taught by one of the teachers in the LEDIMTALI project.

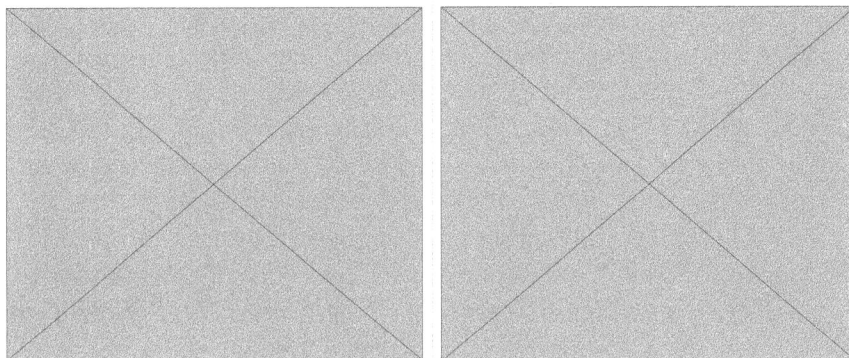

FIGURE 5.3 Two pictures of a learner's work showing an algebraic (on left) and graphical solution (on right) to a quadratic inequality

On the left-hand side, we have a learner's work showing the symbol manipulation or procedures needed to arrive at the truth set for the quadratic inequality, namely in the domain of real ($R[x]$) numbers. This inequality was transformed, as shown in the last line on the left-and right-hand side snapshots in the figure. In the top corner on the right, there is a graphical representation of the transformed inequality. In other words, we have a quadratic function represented symbolically and graphically together with its turning point. This 'sketch graph' is used as a way to find the solution to the inequality. The learner circled on the x-axis of the 'sketch graph' where the truth set can be read off. As stated earlier this 'sketch graph' is over the domain of real ($R[x]$) numbers and is the connector. The main reason for showing the two snapshots is to point out that the teachers know that a quadratic function, graphically represented, has a turning point. They also know the quadratic sequence found in sequences and series, represented in the domain of real ($R[x]$) numbers. In the boundary artefacts, such as the NSC Mathematics examinations questions, quadratic sequences linked to 'series and sequences' are not normally connected to the quadratic function. Also, in the legitimate school mathematics knowledge, questions on quadratic sequence questions are structured in a way that emphasises the constancy of the second difference and the derivation of the symbolic or algebraic quadratic expression, and not so much the turning point of these sequences. As a consequence, there are opportunities for teachers to make mathematical connections between quadratic sequences and quadratic functions.

The above speaks to particular suggestions that might be offered to facilitators (the mathematics teacher educators). Facilitators, therefore, have opportunities to offer teachers ways into designing 'productive practice' activities that can lead to deep progress concerning the institutionalisation of functions in the high-stakes NSC Mathematics examinations questions as a boundary object. This boundary object has gaps that are about the domain of integer ($z[x]$), rational ($Q[x]$), real ($R[x]$) or complex ($C[x]$) numbers and how these appear in the different questions in this boundary object.

My purview as a mathematics teacher educator is informed by a vantage point based on functions and their graphical, symbolic and tabular representations per se. This differs from the separated way that functions appear in this boundary object. What would have happened if I had asked the teachers to extend Table 5.1 themselves? Such an instance of mediation would have meant that the teachers would have had an opportunity to notice the 'turning point' of the quadratic function for themselves. Constant comparison between the two positionalities shows that there was a missed opportunity to 'push' or to expand the teachers' understanding of the quadratic function. The teachers could have been productively engaged if they were asked to do what I did, such as extending the table and noticing a turning point. A paradox is the fact that the teacher institute is formally referred to as designing 'examinations practice' items. This means that the teachers most likely found it necessary to stick to answering and writing the questions on quadratic sequences per se and not on the need to examine 'turning points' peculiar to a graphical and symbolic representation of quadratic functions. There is a benefit in knowing deep mathematical connections between quadratic sequences and quadratic functions when it comes to learners' performance in high-stakes NSC examinations questions.

There was near silence on input about mathematical connections between quadratic series/sequences and the drawing of a 'sketch graph' of the accompanying parabola where the turning point is a maximum. In this incident, I tried to draw attention to the quadratic function over the set of rational numbers and the set of real and/or complex numbers. The near-silence of the teachers can be explained in terms of the strong didactic transposition of the NSC Mathematics content, namely the ways the (quadratic) function concept is institutionalised in the different questions. We do know of teachers from similar education contexts who attempt to work with the function concept as a flexible, unifying underlying concept (Gierdien, 2014). In peculiar ways, with an eye on the high-stakes NSC Mathematics examinations, these teachers 'attempt' to facilitate and to mediate their learners' experiences with 'the function concept.'

Overlaps and 'non-overlaps' between different representations, especially, a tabular representation of the (quadratic) function concept are instances where further research is needed.

At another related level, the above speaks to the possibility of spiral revision and making 'deep progress' which can be allied to 'productive practice' through a focus on functions per se. In examinations practice institutes, the questions are chosen from past papers as found in the high-stakes NSC Mathematics examinations. This approach is driven by the need to attend to the teachers' local concerns referred to earlier on. On the other hand, spiral revision implies that mathematics teacher educators have to learn how to raise awareness about how the mathematics content in terms of functions and how this content is manifested in this high-stakes environment. This content is structured in ways where quadratic functions appear in questions on quadratic sequences/series, as well as in questions where learners have to factorise trinomials and find the roots of, and plot, quadratic equations or inequalities. In other words, there are structuring effects of the boundary objects like the high-stakes NSC Mathematics examinations. Answers to questions on quadratic sequences/series are usually restricted to the domain of rational ($Q[x]$), numbers. Questions on the factorisation and plotting of quadratic functions and inequalities deal with single variable trinomials in the domain of integer ($Z[x]$), rational ($Q[x]$), real ($R[x]$) or complex ($C[x]$) numbers.

In summary, facilitating and mediating means searching for opportunities to make the participating teachers independent observers of mathematical objects, such as functions, that are not just in the high-stakes NSC Mathematics examinations as a boundary object. The context of the designing examinations practice items allowed me to introduce and share my understanding of quadratic functions from a functions vantage point. An important question to ask is: How do I mediate in the sense of introducing learning to the teachers that is not visible in this boundary object?

Concluding remarks ,

This chapter laid some groundwork for thinking about facilitating and mediating on the part of mathematics teacher educators who, through a continuing professional development LEDIMTALI project, work with teachers from socio-economically and historically challenged schools. From the analysis, we saw the usefulness of thinking about different positionalities of the teachers and mathematics teacher educators and the resulting learning opportunities for mathematics teacher educators. To be effective, mathematics teacher educators have to be aware of the mathematical content knowledge that teachers bring to the particular CPD model and how that content is manifested in the context of the high-stakes examinations wherein the teachers work. Depending on the task at hand in the teacher institute or workshop, the teachers will invariably bring to the fore peculiarities related to the didactic transposition of legitimate school mathematics knowledge, which come from their classroom realities. They will be concerned with the ways the mathematics content in the LEDIMTALI project can be used in their classroom teaching. In this case, the in situ or local input of the mathematics teacher educator is crucial and hence the cornerstone intentional teaching model of the present LEDIMTALI project. The mathematics teacher educators have to insert their positionality on this mathematics content in 'intentional' ways and reflect on what happens when they do so. The goal is to learn how to get better at intervening by capitalising on 'moments' to expand teachers' instructional knowledge.

In the current educational context, it becomes necessary to bring to the fore and to study ways of mediating and facilitating different positionalities that arise during and beyond the activities of CPD initiatives. Understanding these different positionalities and how they manifest themselves in CPD helps us to facilitate and mediate professional learning.

References

Boardman, A.G. & Woodruff, A.L. 2004. Teacher change and 'high-stakes' assessments: what happens to professional development? *Teaching and Teacher Education*, 20(6):545-557. https://doi.org/10.1016/j.tate.2004.06.001

Brousseau, G. 1997. *Theory of didactical situations in mathematics: Didactique des mathématiques 1970-1990.* Edited and translated by N. Balacheff, M. Cooper, R. Sutherland & V. Warfield. Dordrecht, The Netherlands: Kluwer Academic Publishers.

Cobb, P.; McClain, K.; Lamberg, T. & Dean, C. 2003. Situating teachers' instructional practices in the institutional setting of the school and school district. *Educational Researcher*, 32(6):13-24. https://doi.org/10.3102/0013189X032006013

Cobb, P.; Zhao, Q. & Dean, C. 2009. Conducting design experiments to support teachers' learning: a reflection from the field, *Journal of the Learning Sciences*, 18(2):165-199. https://doi.org/10.1080/10508400902797933

Freudenthal, H. 1983. *Didactical phenomenology of mathematical structures.* Dordrecht, The Netherlands: Reidel.

Gierdien, M.F. 2014. On the use of spreadsheet algebra programs in the professional development of teachers from selected township high schools. *African Journal of Research in Mathematics, Science and Technology Education*, 18(1):87-99. https://doi.org/10.1080/10288457.2014.884352

Glaser, B.G. & Strauss, A.L. 1967. *Discovery of grounded theory: Strategies for qualitative research.* Chicago: Aldine. https://doi.org/10.1097/00006199-196807000-00014

Harley, K. & Wedekind, V. 2004. Political change, curriculum change and social formation, 1990 to 2002. In: L. Chisholm (ed.), *Changing class, education and social change in post-apartheid South Africa* Cape Town, South Africa: Human Sciences Research Council Press. pp.195-220.

Julie, C. 2002. The activity system of school-teaching mathematics and mathematical modelling. *For the Learning of Mathematics*, 22(3):2-37.

Julie, C. 2012. The primacy of teaching procedures in school mathematics. In: S. Nieuwoudt, D. Laubser & H. Dreyer (eds.), *Proceedings of the 18th Annual National Congress of the Association for Mathematics Association of South African.* Potchefstroom: North-West University.

Julie, C. 2013. Towards a model for intentional teaching for improving achievement in high-stakes mathematics examinations. In: Z. Davis & S. Jaffer (eds.), *Proceedings of the 19th Annual Congress of the Association for Mathematics Education of South Africa*, Vol. 1. Cape Town: AMESA. pp.86-96.

Keitel, C. 2005. Reflections about mathematics education research in South Africa. Afterword. In: R. Vithal, J. Adler, & C. Keitel (eds.), *Researching mathematics education in South Africa. Perspectives, practices and possibilities* Cape Town: Human Sciences Research Council Press. pp.329-344.

Kennedy, M.M. 2016. How does professional development improve teaching? *Review of Educational Research*, 86(4):945-980. https://doi.org/10.3102/0034654315626800

Kvale, S. 1993. Examinations re-examined: Certification for students or certification of knowledge? In: J. Lave & S. Chaiklin (eds.), *Understanding practice: Perspectives on activity and context* Cambridge, UK: Cambridge University Press. pp.215-240. https://doi.org/10.1017/CBO9780511625510.009

Labaree, D. 2011. Targeting teachers. *Dissent*, 58(3):9-11. https://doi.org/10.1353/dss.2011.0068

Marick, B. 2001 Boundary objects. Available: http://bit.ly/2LkCJGE (Accessed 3 February 2014).

Milner, R. 2007. Race, culture, and researcher positionality: Working through dangers seen, unseen and unforeseen, *Educational Researcher*, 36(7):388-400. https://doi.org/10.3102/0013189X07309471

Nyikahadzoyi, M.R. 2006. *Prospective Zimbabwean: A level mathematics teachers' knowledge of the concept of a function.* Doctoral dissertation, University of the Western Cape, Bellville, South Africa.

Poekert, P. 2011. The pedagogy of facilitation: Teacher inquiry as professional development in a Florida elementary school. *Professional Development in Education*, 37(1):19-38. https://doi.org/10.1080/19415251003737309

Selter, C. 1996. Doing mathematics whilst practising skills. In: C. van der Boer & M. Dolk (eds.), *Modellen, meting en meetkunde: Paradigma's van adaptief onderwisj.* Utrecht: Panama/HvU & Freudenthal Institute. pp.31-43.

Sfard, A. 1998. The many faces of mathematics: do mathematicians and researchers in mathematics education speak about the same thing? In: A. Sierpinska & J. Kilpatrick (eds.), *Mathematics education as a research domain: A search for identity* (Vol. 2). Dordrecht, The Netherlands: Kluwer Academic Publishers. pp.491-511. https://doi.org/10.1007/978-94-011-5196-2_18

Smith, M.L. 1991. Put to the test: The effects of external testing on teachers. *Educational Researcher,* 20(5):8-11. https://doi.org/10.3102/0013189X020005008

Spaull, N. 2013. *South Africa's education crisis: The quality of education in South Africa 1994-2011.* Report commissioned by the Centre for Development and Enterprise (CDE).

Star, S.L. 1988. The structure of ill-structured solutions: Boundary objects and heterogeneous distributed problem solving. In: M. Huhns & L. Gasser (eds.), *Readings in distributed artificial intelligence.* Menlo Park, CA: Kaufman. https://doi.org/10.1016/B978-1-55860-092-8.50006-X

Star, S.L. 2010. This is not a boundary object: Reflections on the origin of a concept. *Science, Technology & Human Values*, 35:601-617. https://doi.org/10.1177/0162243910377624

Star, S.L. & Griesemer, J.R. 1989. Institutional ecology, 'translations' and boundary objects: Amateurs and professionals in Berkeley's Museum of Vertebrate Zoology. *Social Studies of Science*, 19:387-420. https://doi.org/10.1177/030631289019003001

Stein, M.K. & Coburn, C. 2007. *Architectures for learning: A comparative analysis of two urban school districts.* Seattle, WA: University of Washington, Center for the Study of Teaching and Policy.

Teese, R. 2000. *Academic success and social power: examinations and inequality.* Melbourne: Melbourne University Press.

Wall, D. 2000. The impact of high-stakes testing on teaching and learning: can this be predicted or controlled? *System*, 28(4):499-509. https://doi.org/10.1016/S0346-251X(00)00035-X

Watson, A. 2008. School mathematics as a special kind of mathematics, *For the Learning of Mathematics*, 28(3):3-7.

Watson, A. & De Geest, E. 2005. Principled teaching for deep progress: improving mathematical learning beyond methods and materials. *Educational Studies in Mathematics*, 58(2):209-234. https://doi.org/10.1007/s10649-005-2756-x

Watson, A.; De Geest, E. & Prestage, S. 2003. *Deep progress in mathematics.* Oxford, UK: University of Oxford, Department of Educational Studies.

Wenger, E. 1998. *Communities of practice: Learning, meaning, and identity.* New York: Cambridge University Press. https://doi.org/10.1017/CBO9780511803932

Wenger, E. (2000). Communities of practice and social learning systems. *Organization*, 7(2):225-246. https://doi.org/10.1177/135050840072002

|06|

PRACTICING TEACHERS AND THE DEVELOPMENT OF MATHEMATICAL MODELLING COMPETENCIES THROUGH MATHEMATICAL MODELLING AS CONTENT

Cyril Julie

Introduction

Mathematical modelling is a feature of the school mathematics curriculum in many countries, including Denmark, Australia and the Netherlands. The South African school mathematics curriculum stipulates that mathematical modelling is an anchoring feature of the school mathematics curriculum. CAPS states:

> Mathematical modelling is an important focal point of the curriculum. Real-life problems should be incorporated into all sections whenever appropriate. Examples used should be realistic and not contrived. Contextual problems should include issues relating to health, social, economic, cultural, scientific, political and environmental issues whenever possible. (Department of Basic Education, DBE, 2011:8)

This statement appears in CAPS for all phases of schooling in South Africa. The curriculum does not clearly state how mathematical modelling should be dealt with but does say that real-life problems should be realistic and not contrived. This suggests that modelling problems should be approached from the perspective similar or near-similar to that of the practice of expert mathematical modellers. A feature of the problems for which mathematical models need to be developed is that they are generally messy. A dilemma associated with the espoused prominence of mathematical modelling in the school mathematics curriculum is that the mathematics teaching force tasked with implementing mathematical modelling had limited (or even no) substantive encounters with mathematical modelling during their pre-service teacher education. Pre-service teacher education mathematics courses at all levels (Foundation, Intermediate, Senior and Further Education and Training Phases) are generally anchored around pure mathematics. Consequently, continuing professional development providers for mathematics teachers have a responsibility to induct practicing teachers into mathematical modelling and allow them to experience mathematical modelling and the applications of mathematics in some substantive manner, particularised for the phases they are teaching. This will at least sensitise teachers to the intricacies involved in mathematical modelling as espoused in the curriculum.

This chapter serves as an introduction to mathematical modelling of practicing Grades 7-12 mathematics teachers with limited or no experience of mathematical modelling. It is well known that teachers should experience engaging with mathematics in the same way it is desired that learners should. Purportedly, this would provide teachers with some experiential base about doing mathematics from which to draw. The premise is no different for mathematical modelling. This is in line with Niss, Blum and Galbraith (2007:7), who suggest that practicing teachers "need opportunities to develop [the] capacity [to do mathematical modelling] … through regular in-service activities of professional development" (Niss, et al., 2007:7).

There is an emerging body of literature dealing with practicing teachers and mathematical modelling. This is evident from the various books, such as the one by Haines, Galbraith, Blum and Khan (2007), resulting from the deliberations during the biennial conferences of the International Community of Teachers of Mathematical Modelling and Applications (ICTMA) and other related journals.[1] The corpus, however, focuses primarily on issues such as practicing teachers' perceptions of the modelling process, their beliefs of mathematical modelling, pedagogy and their implementation of applications and modelling in their classrooms. Research regarding practicing teachers' engagement with modelling as content is limited. In the ICTMA publications from 2007-2015, very few studies explicitly focus on practicing teachers' doing of mathematical modelling. One of the few is the study by Ng (2013) regarding teacher readiness in mathematical modelling.

---

[1] See, for example, Blomhøj and Kjeldse (2006).

This chapter addresses the issue and focuses the ways of working when practicing teachers experience mathematical modelling for the first time from a modelling as content perspective.

Research framework

Julie and Mudaly (2007) and Julie (2002) demarcate two distinct approaches to mathematical modelling in teaching situations. These are modelling as a vehicle and modelling as content. The former is described as:

> embedding mathematics in contexts … not [for] the construction of mathematical models per se, but rather the use of contexts and mathematical models as a mechanism for learning of mathematical concepts, procedures, conjectures and, at times, developing context-driven justifications for obtained conjectures. (Julie & Mudaly, 2007:504).

This is by far the major approach on how pupils and most students experience mathematical modelling. The most elementary way of this experiencing is in word problems and the key to this approach is that the problems pupils deal with have one fixed answer although there can be different ways of reaching the answer.

The second approach is defined as follows:

> Mathematical modelling as content entails the construction of mathematical models for natural and social phenomena without the prescription that certain mathematical concepts or procedures should be the outcome of the model-building process. It also entails the scrutiny, dissection, critique, extension and adaptation of existing models with the view to come to grips with the underlying mechanisms of mathematical model construction. (Julie, 2002:2)

An intention of the "mathematical modelling as content" approach is that in teaching, participants should experience modelling in a near similar way that adept modellers experience it in practice. Julie (2002, 2015) characterises this practice in terms of a requester/client and model builder/constructor configuration. As is obvious, this configuration encompasses two parties – the one requesting a model for a particular purpose and the other a provider of the requested service. In some instances, the requester and modeller can be the same person or a group, as is the case where models are developed for phenomena of interest in an extra-mathematical disciplinary domain. In this regard, Thompson (2015) gives a popular account of how a group comprising a mathematician and two physicists developed a 'Big Rip' theory or a model related to the expansion rate of the universe.

Mathematical modelling is generally viewed as the development of a mathematical representation of an extra-mathematical phenomenon to describe, predict or make prescriptions regarding the issue of import. The process of mathematical modelling is normally given as steps or stages. These stages present an ideal-typical rendition of the mathematical modelling process as found in various articles related to mathematical modelling.[2] Summarily these 'steps or stages' are the translation of a real-life problem to a mathematical one, the solution of the derived mathematical problem, the validation of the mathematical results against the real-life situation and the declaration of the model as the current best mathematical representation for the situation. Although the 'steps or stages' are used to characterise the process, there is a consensus that the mathematical modelling process is not linear and that there is much movement between the different stages when a model is developed.

Regarding the practice of mathematical modelling by expert/mature mathematical modellers, Paul Davis captures the way they encounter problems they must resolve as follows "Many problems are posed to industrial mathematicians by colleagues in other disciplines, who may not yet understand the real problems they face. Good problems need not be elegant, new, or well posed, just necessary to corporate welfare" (Davis, 1991:3).

From the above, it can be extracted that the mathematical modelling way of working in adept practice comprises of a vague formulation of an actual real-life problem, but the outcome is known – what the client wants! A process of discussions and consultations for consensus-generation between the developer and requester the problem refines the problem (Julie, 2004). Following the reaching of consensus, the model-developer constructs the mathematical model, which is delivered and explained to the client. There might be requests for further fine-tuning of the model to better suit the purposes of the client. After completion of these activities, the model-builder stores the model for adapted use for near-analogous future requests or further development as new insights and demands emerge. Concerning the latter, Sen (1976:228), for example, presents the evolution of models to determine the poverty index and ends up with one, which he claims "have quite a simple interpretation".

Mathematical competencies are derived from the ideal-typical cyclical representation of the mathematical modelling process encapsulated by the steps

---

above. The Organisation for Economic Co-operation and Development (OECD) defines 'competency' as follows:

> [T]he ability to meet individual or social demands successfully, or to carry out an activity or task. This external, demand-oriented, or functional approach has the advantage of placing at the forefront the personal and social demands facing individuals. This demand-oriented definition needs to be complemented by a conceptualization of competencies as internal mental structures – in the sense of abilities, capacities or dispositions embedded in the individual.
>
> Each competence is built on a combination of interrelated cognitive and practical skills, knowledge (including tacit knowledge), motivation, value orientation, attitudes, emotions, and other social and behavioural components that together can be mobilized for effective action. (OECD, 2002:8)

A similar definition of competency is normally used for mathematical modelling. Blomhøj and Jensen (2003:126), for example, define competency as "someone's insightful readiness to act in a way that meets the challenges of a given situation". Katja Maaß proposes a similar definition of competency. She specialises the definition for mathematical modelling competencies as "Modelling competencies include skills and abilities to perform modelling processes appropriately and goal-oriented as well as the willingness to put these into action" (Maaß, 2006:117). Maaß arrives at her definition of modelling competencies through widely available representations of the mathematical modelling process, however, she also develops sub-competencies for the competencies identified in the mathematical modelling process. Sub-competencies are developed for each of the following competencies to:

> understand the real problem and … set up a model based on reality … to set up a mathematical model from the real model… solve mathematical questions within this mathematical model… interpret mathematical results in a real situation [and] validate the solution. (Maaß, 2006:116-117)

Maaß's definition underpinned the study for the research reported here, however, the competencies inherent in Giordano and Weir's (1985) representation of the mathematical modelling process are brought into the picture. These are:

1. identification of the problem;
2. making necessary assumptions comprising of (a) identification of variables, and (b) determination of interrelationships between variables;
3. solving the model;
4. verification of the model through questions such as (a) Does it address the problem? (b) Does it make common sense? (c) Test it with real-world data;
5. Implementing the model; and
6. maintaining the model. (Giordano & Weir, 1985:35-37).

The similarities between Giordano & Weir's steps for mathematical modelling and Maaß's competencies are clearly discernible. The choice of the steps of Giordano and Weir (1985) is more a question of convenience since it is one with which the participants in this study were more conversant with due to its widespread appearance in South African curriculum documents.[3]

The participants, setting and problem

Ten high school teachers, one primary school teacher and a mathematics curriculum advisor – a former high school mathematics teacher – were involved in this study. The mathematics courses they completed during their teacher education varied, with the lowest being normal school manipulative algebra including graphs of quadratic functions and the highest being courses at post-graduate mathematics level which focused on abstract algebra and logic. Each teacher had over ten years' experience in teaching school mathematics. Barring the primary school teacher, they taught mostly at the upper levels of high school and the curriculum included differential calculus focusing on the derivatives of elementary algebraic functions using 'limits'. Their experience regarding mathematical modelling was restricted to applying mathematics to contextual situations linked to the mathematical topics immediately after these topics were taught or those appearing in examinations set by external examining bodies. Figure 6.1, for example, is a problem from a 2014 Independent Examination Board (IEB) Mathematics Paper 1. It is the final sub-question of a question related to sequences.

An athlete runs 20 km on a certain Monday. Thereafter, he increases the distance by 10% every day.

Calculate the number of kilometres he ran:

1. on the following Saturday
2. altogether over the 6 days

Figure 6.1   Contextual situation problem set in an examination
Source   *IEB (2014:3)*

The setting was a residential workshop, which teachers attended from a Friday afternoon to Sunday midday. The course was a continuing professional development short course of 16 hours of scheduled engagement time – four hours on Friday and Sunday and eight hours on Saturday. The focus of the course was an introduction to elementary mathematical modelling. The course content

---

3   See, for example, Olivier (n.d.).

was (a) an introduction to modelling; (b) the analysis and development of an elementary ranking model; and (c) the development of a model for a situation selected from a set of prescribed situations, namely:

∅   parking lot design;

∅   control of an elephant herd in a national park in South Africa;

∅   a ranking system for South African universities based on publications output; and

∅   the placement of emergency service facilities and a situation analogous to the one dealt with in part (a).

This chapter focuses on part (a) of the course. This part of the course consisted of:

∅   a modelling activity followed by a lecture on the modelling cycle and modelling competencies;

∅   the development and report on the model for the situation given below; and

∅   commentary on the model by an outsider who was knowledgeable about mathematical modelling and the associated competencies.

The course generally followed an immersion teaching approach. This approach is similar to the holistic approach (Blomhoj & Jensen, 2003) and is characterised by the way students encounter the issue to be modelled or the way the problem is formulated. In this approach, an environment is provided which near-simulates the way applied mathematicians and mathematical modellers of mathematics in industry, business, government, and so forth, encounter a situation to be modelled. This ensured that the participating teachers were put in a situation to experience mathematical modelling as content. From the requester-model developer perspective, the course leader played the role of the requester of the model and the participating teachers that of mathematical modellers. As is the case in mature practice, the course leader as requester provided, upon request, clarifications of what is expected of the final model.

The immersion teaching approach is normally preceded by a discussion on the applications of mathematics and mathematical modelling where some versions of the ideal-typical processes of modelling and associated competencies are presented. This is done before participants are requested to develop a model for a situation. For this study, this approach was inverted in that the participants were requested to start the development of a model before any exposition of the mathematical competencies was done by the facilitator. Participants were presented with the contextual situation given in Figure 6.2 and they were requested to start developing the model.

What space is occupied by the tree?

FIGURE 6.2   The modelling situation and the problem formulation

The participating teachers worked collectively and collaboratively on the problem. The moving to the exposition of the ideal-typical mathematical process commenced when the facilitator used his judgement that the participants had engaged sufficiently and were nearing the construction of a model.

Data collection and analysis

The sessions where teachers worked in the teaching venue were video recorded. For the study reported here, the data analysed consisted of approximately five hours of video footage. The constant-comparative analysis (Corbin & Strauss, 2008) method was used to create categories. The derived categories were interrogated to identify the competencies that became visible through the real-time work of the participants.

Since the course was residential it is reasonable to assume that the teachers had some form of interaction with the issues of the course during non-scheduled times, such as dinner, breakfast, and so forth. The researcher was not privy to the relevant discussions during these interactions.

Results and discussion

This section presents the categories that were developed through the analysis process. The analysis rendered five broad categories, which include discussions of the competencies that are ostensibly at play in the way of working of the practicing teachers. These categories are presented in the five sub-sections that follow.

School mathematics content knowledge as a solution-seeking driver

After studying the task, clarification of what the task entails commenced by one of the participants asking, "*What is the question here*?" Another teacher responded, '*Floor space*'. This ('floor space') response is indicative of the enculturated school mathematics that is drawn on. In school mathematics, 'area' is defined as "the amount of space that a flat surface or shape covers" and 'floor' are used in application problems related to an area. A typical problem in school mathematics, for example, is "For the floor of a room 4,5 m by 3,75 m, find the area of the floor in m². (Bowie, et al., 2012:119). Hence, it is reasonable to assume that at this stage 'space' is taken as 'area'. The follow-up response is "What about the roots under the ground? It is also part of the space." This indicates that there is a possible awareness that there is more at play than just what is visibly displayed. This points in the direction of a possible assumption but it is at a very implicit and vague level at this stage. The trend of the conversation is changed by a teacher asking, "What's the formula for the [space of] the tree?" The dominating effect of the enculturated school mathematics is again noticeable where when area is dealt with in school mathematics it is done in conjunction with given formulae such as the area of a triangle $= \frac{1}{2}$ base x height.

Structuring by school mathematics for pursuing an entry point and Giordano and Weir's (1985) identification of the problem continued. Another teacher said: "What shape is this?" [Circles top of the tree with his pen]. If a formula is at play the kind of shape must be known. Implicitly, some known model 'kind of shape' is alluded to but still in the frame of school mathematics.

The need for dimensions

A shift of focus to the elements needed for calculation of area or space is introduced as evident from the interchange between participants:

Teacher H (TH)   You can only occupy space if there are values for the space. If there is no values for the space, then it is not occupying space. This is x and this is y [Indicates vertical and horizontal borders of photo].

Teacher S (TS)   What space …

Teacher H (TH)   …not how much space …

Teacher M (TM)   What space …

TS   So, so now … [others laughing] …a … space …

Almost chorus-like response]: What space … Not adequate information. Not enough data … No space no values.

The excerpt indicates that the absence of 'values' caused some concern. It appears that the participants felt that they could only proceed if concrete dimensions were provided. This is linked to the above category of school mathematics as a driver. In their experience of teaching shape and space – a defined topic in the school mathematics curriculum – the problems are always accompanied by given dimensions, either as numerical values, variables or both, as is evident in Figure 6.3.

A rain gauge is in the shape of a cone. Water flows into the gauge. The height of the water is h cm when the radius is r cm. The angle between the cone edge and the radius is 60, as shown in the diagram below. Determine r in terms of h. Leave your answer in surd form.

(The formulae $V = \pi r^2 h$, $V = lbh$, $V = \frac{1}{3}\pi r^2 h$ are provided.)

FIGURE 6.3   Example of a problem related to shape and space

Source   *DBE (2015:9)*

Also evident from the excerpt is that the use of variables is not yet part of the deliberations of the participants.

Although the focus was on the absence of dimensions, there was an attempt to get a way out of being 'stuck' by considering 'reality' at some philosophical level. A teacher trying to steer the discussion in this direction said:

> You can see that the space occupied by the tree needs to go. You can look at maybe the value of the space because it's the value to the space … then the tree will occupy that space … if there is no value then the tree does not occupy the space

Another teacher immediately retorted by saying: *"But the tree is there"*, so countering the attempt to move the situation to a non-existent one by the former teacher.

Absence of dimensions – considering a comparison strategy

The participants were now fixated on 'values' that are needed. They reverted to a kind of comparison strategy using what they can 'see'. The deliberations proceeded as follows:

Teacher MS (TMS)	You can say the space is half here… [Finger-drawing on the trees to indicate that he means the region covered by the different trees visible in the photograph.]
Teacher H (TH)	You can say that the tree occupies a third of the space…
TMS [insists]	The tree is half the space…because it [consist of…]
Teacher J (TJ)	[We] should all agree on a third of the photo.

TJ gets some support by a chorus-like answer:

TJ	if correct, we are all correct and if wrong all wrong

This support is strengthened by TS who says:

Teacher S (TS)	If we disagree … [we] all have our own ideas and maybe one might be right. Let's write them all.
TH	A third is suggested. [On his problem sheet he indicates a question mark.]
Teacher W (TW)	Too obvious, hence question mark – sometimes [the] obvious is correct. [Referred to a half that was offered.] When the whole page [is used] as reference – about one-tenth.

The use of the sheet, the images on it and the attempts to compare the regions covered by the images with the entire sheet point in the direction that the quest for actual 'values', was abandoned. Also, the gesticulations of tracing on the sheet and focusing on "part of the whole sheet" show some sense that the essence of the problem to be addressed is coming into focus.

Initiation into problem specification

The deliberations started to move in the direction of settlement of the problem to be addressed by bringing into sharper focus elements for the selection of components of the model and the possible mathematical area to pursue. The conversation continues:

TS	Which tree are we talking about? The one in front or all the trees? It is such an open question … there won't even be a memo for it…
Teacher Wh (TWh)	What about the roots under the ground? It is also part of the space …
TS	You must give clear instructions … because otherwise all the nonsense come out … [referring to learners].
Teacher MN (TMN)	So the space is the space on this photo…
Teacher MD (TMD)	On the ground or in the atmosphere?
TWh	It is not area. It is volume
TJ	In the exam for calculating the volume of a box a 2D representation is provided. What is seen – only the flat picture on the sheet or the box?
TH	2D with no depth coming out. In real-life there is depth. Modelling is applied to real-life. This is a picture of a real tree. Could not bring the tree here. It's 3D; it's got volume, cylinders, roots, leave … Lots of ways you can work with leaves – triangles, rectangle, triangle again. Physically take the tree and use dimensions to calculate its volume. Compare to do a calculation. Make a comparison to link it to what the size should be. Again find everything else and subtract and the what's left is the volume.

The excerpt shows that the mathematical object, area or volume, that will be considered is coming more sharply into focus. Linked to the stages for mathematical modelling it can be claimed that a point near to proceeding from the mathematical formulation of the problem to the intra-mathematical domain is being implicitly approached.

When one of the participants uttered "Again find everything else and subtract and the what's left is the volume", the facilitator intervened and suggested that a representation in the form of a drawing be made. The participant who made the utterance further explained what he thought should be done by representing his view as in Figure 6.4.

FIGURE 6.4   Drawing of the "what's left is the volume"

Note   *Observable in the drawing is that a variable – H – is emerging.*

The group then went in circles debating whether area or volume should be the focus. The facilitator intervened by telling them to split into 'area' and 'volume' groups. Two participants chose to consider an area model and the rest a volume model. At this juncture after nearly two hours of grappling to settle on the problem to be focused on the construction of the mathematical model was suspended and the lecture on mathematical modelling mentioned previously was presented.

Construction of the model

At the start of the next session (the next day), one member of the '2D'-group had to leave the group. The other member of this group changed his mind and joined the '3D'-group. The work session commenced with the facilitator asking: "*Did anyone develop a model?*" There was no specific response to the question but a suggestion about filling a glass with water and measuring it to get to grips with 'space'. To keep them on track regarding the development of a model the facilitator asked: "*Is it possible to develop a model?*" The response to this question was positive with utterances such as "*we can say one part of the tree looks like a cylinder … we can develop one way. We can look at the way we find the volume of a cylinder … and I think it is a sphere."*

The participants were now asked to start working on developing the model. One of the participants asked about the "*leaves and stuff*" to which the response from another participant reminded him that "*when we make a model, we make an assumption. You assume that it looks like a sphere …*" A participant, TMD, was called to start working publicly on the newsprint which was used as a recording device. The group felt that they wanted to first work on it collectively as a group. They put tables together to form a rectangular surface that the entire group could participate at one workspace. They started on the problem as indicated in the following excerpt:

Teacher AS (TAS)	Draw a rough sketch
TS	You are only talking about this bottom part now …
TMD	No, no … the tree … just a rough sketch …
TAS	That's the bottom of the tree. Now the top of the tree.
TMD	The top of the tree. What is the shape of the top of the tree?
TS	It might be triangular … It looks like triangular, maybe triangle.
TJ	I would say triangle.
TMD	Uh, uh …
TMS	It says what space does it occupy, so …
TAS	So the shape of the tree …
TMS	Let's see the shape of the tree. Try to make it three dimensional.
TS	We try to get the top part.
Some members simultaneously	Triangle!
TS	Triangle is two-dimensional.
TWh	Can't it be a cone?
TS and Teacher Dun (TDun)	Cone, cone …

Some members simultaneously	Can't be …
TW	No, no … but the cone is 3D.

Triangle is mentioned and discarded because it is not 3D.

TS	The best is the cone …
TAS	[Writes and says]: Volume of the cone plus volume of the cylinder …
TMD	We use 'r' for our radius and 'h' for our height. Smaller 'r' and smaller 'h'.

The above description and accompanying excerpt demonstrate that the identification and unravelling of the problem is completed (cf. the 'first step' in Giordano and Weir, 1985). Although the deliberations veered in the direction of a '2D'-model, the intervention of one of the participants in the form of a question *"Can't it be a cone?"* steered the course of discussion into accepting that it must be a cone because of the '3D'-approach that was adopted. The teacher who intervened with the cone suggestion was a student doing an advanced qualification in mathematics at the African Institute for the Mathematical Sciences (AIMS). It is surmised that his current involvement in advanced mathematics triggered the idea of solids of revolution and that a cone can be pictured as being generated by rotating a right-angled triangle line through some axis of revolution. Further competencies evident from the group are the start of the declaration of assumptions, the identification of variables and the splitting of the problem into two sub-problems.

The sorting out of a model now shifts to the other sub-problem called "the bottom of the tree". The following excerpt presents the discussion.

TAS	Now for the bottom of the tree … what is the bottom of the tree? Or is it spherical?
TW	The bottom part. What about the roots?
TS	We are not gonna worry about the roots. We are only gonna work with the top part.
TWh	We assume that we are just gonna deal with what is visible.
Group more or less simultaneously	We deal only with what is visible

The member who joined from the 2D-group was again trying to get the group to consider a triangle. Some debate and admonishment about him wanting to do a 2D-model and then joining the group, so he has to abide by the 3D-notion or "bring his 2D-model to the table". The facilitator intervened by asking the group whether they have ever seen a 2D-tree in 'real-life'.

TMS	We can use those measurements on the 2D drawing to also determine the volume.
TMD	Look basically what we are doing in modelling … We are not interested in figures … You are developing a mathematical model.
TAS	There is a model, but it is based on assumptions …
TS	Assumptions yes. For that one at the top, we say it is a cone, we are assuming it is a cone … That will take into account the … the top of the tree is …
TMD	You will devise a way to get the actual size … but the bottom line is we can't go to the actual figures before we get … because we still have to model. Now we are going to find the model because once we get the formula then it would be easy for us to plug it in the formula for …
TMS	any tree …
TMD	Any tree yah …
TAS	But the shape of this one [leave-covered upper part] is a cone and the places between the leaves compensated by the places where it is less dense.
TMD	So we say we find the volume of this … plus the volume of this. [The indexical 'this' referring to the leave-covered upper part in the first instance and the trunk in the second instance.]
TS	Plus …
TAS	No we don't need that … we said we only look above the ground.
TMD	But we are assuming that this … [above the ground is circular]
TW	You must just state that …
TS	So now we have the formula for a cone … now we must do this part [ostensibly the trunk].

Chorus-like	The cylinder
TMS	[Reads from his mobile phone. Possibly Googled the formula for a cylinder] 2, pi, r, times height …
TDu	What is 2, pi, r …
TW	The area must be taken …
Chorus	pi, r, squared …
TW	But then we take that radius in the centre … it is getting a little thinner as we move up … there [refers to the variable thickness of the trunk as the height of the trunk nears the start of the 'conic' part.
TMS	Then we must use the diameter.
TW	No, no, for the area it will be the radius …
TAS	But now remember …
TW	What I say we take the radius of the middle of the trunk … just to get an estimation. The bottom will be a little bit wider and the top will be smaller, but we take it in the middle. You can take the radius there …
TMS	So what is that now. That will be the model … So for any other tree that will be the model.
TAS	For any tree if the shape is the same …

Having sorted out the 'model' they would present, one member was designated to present their initial ideas and model on the newsprint. She started her explanation by drawing the representation of the tree (Figure 6.5 – Figures 6.5, 6.6 and 6.7 are reproduced from the original newsprint sheet).

The radius of the trunk is constant. No space between branches and leaves.

We use only the visible part of the tree which excludes the roots.

The top part is the cone.

And the trunk is cylindrical from around the first branch of the tree.

FIGURE 6.5   Completed working representation of the tree

With the emergence of the diagram, she stressed that they had assumed certain things:

Presenter   We assumed as we go down there is this trunk … is a cylinder [contributed by another teacher] … we assumed it is very difficult for us to see whether it is circular because the roots are going in many directions.

One of the teachers interrupted and said:

We must take into consideration that the radius of the trunk changes as we go up. We are working with estimations …

Various other members of the group tried to convince the dissenting member by referring to the assumptions that were being made. The facilitator intervened by informing the group of the importance of the assumptions they were making. The recorder started to write the assumptions on the newsprint and eventually ended up with the preliminaries for the model as depicted in Figure 6.6.

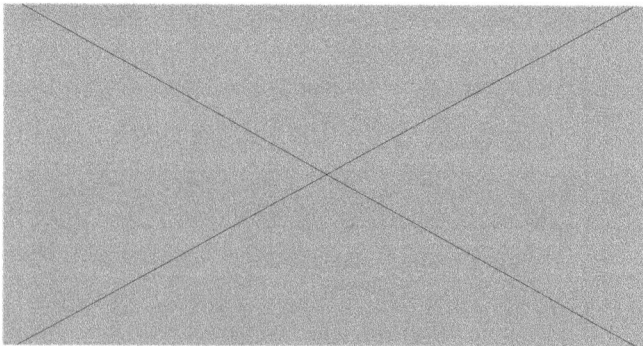

Figure 6.6   The represented preliminary model with the assumptions

As can be observed, the first assumption is linked to the issue raised about the uniform thickness of the trunk. This is followed by other members recalling other assumptions that emerged during their deliberations. The facilitator drew their attention to the kinds of solids they settled on and that they also had to include those as assumptions. These were added and another member suggested that they should also indicate that the trunk is measured from the "ground to the first branches of the tree".

The dissenting member now wanted to know whether the distance the photographer stood from the tree should also be included as an assumption saying that the size of the tree on the photograph will change dependent on the distance of the photographer from the tree. Various other members tried to clarify what they were busy with. Utterances, such as, "we are not working with numbers, we are doing it for any tree and we not busy with scaling", were made. The facilitator intervened again by telling them the focus of the task is to find the space occupied by the tree in 'real-life'. Another participant mentioned, "Your variables will take care of the height." The facilitator then drew their attention to a distinction between a model for a particular object and a general one for a tree with the shape of the one given in the photograph. The recorder wanted to move to write down the variables to be used and one participant wanted to know whether using things such as r is also not an assumption. The facilitator intervened by informing the group that the variables are mere declarations. The recorder continued to complete the model from which a preliminary 'rough' poster was designed. The preliminary poster was converted to one, with a section given in Figure 6.7, the group presented to an external expert on mathematical modelling the following day.

Model

The volume of the tree = volume of cone + volume of cylinder

$$= \tfrac{1}{3}\,\pi R^2 H + \pi r^2 h$$

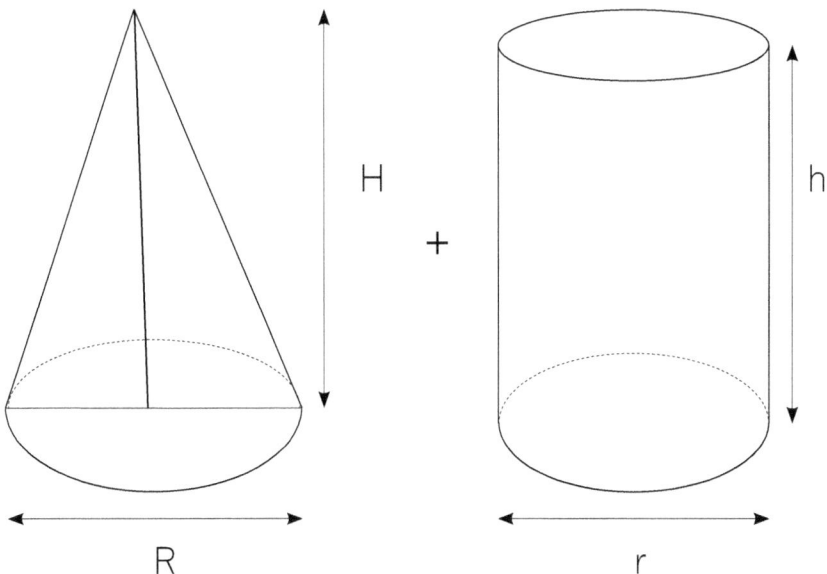

FIGURE 6.7   Section of poster of 'tree model'

The preliminary poster was converted to one the group presented to an external expert on mathematical modelling the following day. He commented favourably on their work but drew their attention of the specificity of the model for a particular tree type and that they had to consider generalising their model for any tree.

Concluding discussion

The thrust of this chapter is on the mathematical modelling competencies that were triggered and emerged while practicing teachers were engaged in the construction of a model embedded in the perspective of mathematical modelling as content. What became clear in the exercise was that the competency of clarifying the problem to be tackled took up a major portion of the model construction time. This competency has been "identified as among the most difficult phases of the entire [modelling] process" (Galbraith, Stillman, Brown & Edwards, 2007:131). Biembengut and Hein (2010:487) concur with this assertion regarding the difficulties teachers experience when confronted with contextual situations. In actual mathematical modelling practice, a component of this phase is the consultation between the model-requester and model-developer. Caron and Belair (2007:122) characterise this as follows: "In most real-world situations where mathematics is applied, it is very much in dialog with non-mathematicians [to gain] a clearer understanding of the situation." Furthermore, according to the blockages framework developed by Galbraith and Stillman (2006:147) during the clarification phase the "ACTING OUT, SIMULATING, REPRESENTING, DISCUSSING PROBLEM SITUATION" (capitilisation in original) was focused on making a decision regarding the *'mathematical entity'* to consider (Galbraith et al., 2007:137. Italics in original). The way the participants grappled to decide on an ultimate mathematical entity shows that the *'level of intensity'* of the blockage was high (Stillman, Brown & Galbraith, 2010:393. It seems, therefore, that practicing teachers as novice modellers encounter blockages similar to those of pupils. Ng (2013:341) ascribes these blockages displayed by teachers as novice modellers approaching modelling tasks from a didactical vantage point rooted in the mathematics they were teaching. Although, as the data in the example demonstrate, this is a truism, I contend that it is also the dominant experience of mathematical modelling as a vehicle that causes teachers to find this phase challenging in the sense of "if you have nothing to start with start with what you've got". Allowing practicing teachers to experience this aspect first-hand has the potential value of sensitising teachers to tolerance when learners grapple with this phase of the modelling process.

Noticeable also in the workshop that despite this issue being made clear during the accompanying lecture on modelling there was no discussion on the use of the model in real-life situations. It is well known that a mathematical model always has a purpose, which could be descriptive, predictive or prescriptive (Davis & Hersh, 1988). The ignoring of the possible uses of the model had the consequence of teachers not "Considering Real-World implications of mathematical results" (Galbraith & Stillman, 2006:147). It is surmised that the dominance of thinking and working along school mathematics lines seems to structure teachers' focus. In so doing, it results in teachers ignoring certain aspects of the modelling process. It is suggested that this issue has to be addressed in courses, of whatever nature, for capacitating teachers to do mathematical modelling. More explicit visible making of this aspect using the client/model developer perspective, preferably with scenarios based on data obtained from actual initial and on-going discussions and deliberations between client and model developer to clarify the problems, offers possibilities to deal with this issue.

A further exertion of agency by the dominance of thinking and working along school mathematics lines is the decision that when a model is designed and the 'answer' is found then there is no need for further refinement of the model, the investigation of a possibility of a different model with higher fidelity or the extension of the model beyond the local. In other words, most of the issues referred to by Galbraith and Stillman (2006:147) under their "REAL WORLD MEANING OF SOLUTION → REVISE MODEL OR ACCEPT SOLUTION" did not trigger. In particular, the "REAL WORLD MEANING OF SOLUTION", as alluded to above did not surface. This was despite the explicit addressing of these issues during the lecture. It was only after the commentator, upon the presentation of their model to him, drew the group's attention to the local nature of their model that one participant did further work on globalising the model, which he presented at a subsequent professional development activity.

There is general agreement that "teachers should have authentic experiences that involve modelling" (De Oliveira & Barbosa, 2010:511). The study revealed the blockages teachers have with dealing with mathematical modelling as content. It opens up further discussion on the directions to be pursued to engage practicing teachers as modellers in continuing professional development initiatives. This, I daresay, should also be the case for pre-service teachers. However, it will not be easy to appropriate these experiences for pedagogical actions. This is, in part, due to mathematical modelling as content being removed from and different to modelling approaches prioritised in the school mathematics distributed through textbooks and examinations, which are the main sources teachers consult and use for their teaching. Nevertheless, continuing professional development initiatives and initial teacher education programmes should also provide practicing teachers with experiences of mathematical work that does not seemingly have immediate relevance to their practice. This widening of perspectives of mathematical work might then, as the difference between the intended and implemented curriculum decreases, be utilised to engage teachers in pedagogical activities linked to mathematical modelling as content.

References

Biembengut, M.S. & Hein, N. 2010. Mathematical modelling: Implications for teaching. In: R. Lesh, P.L. Galbraith, C.R. Haines & A. Hurford (eds.), *Modeling students' mathematical modeling competencies.* New York: Springer. pp.481-490. https://doi.org/10.1007/978-1-4419-0561-1_41

Blomhøj, M. & Jensen, T. 2003. Developing mathematical modelling competence: conceptual clarification and educational planning. *Teaching Mathematics and its Applications,* 22(3):123-139. https://doi.org/10.1093/teamat/22.3.123

Blomhøj, M & Kjeldse, T.H. 2006. Teaching mathematical modelling through project work – Experiences from an in-service course for upper secondary teachers. *ZDM,* 38(2):163-177. https://doi.org/10.1007/BF02655887

Blum, W. & Leiß, D. 2006. 'Filling up' – the problem of independence-preserving teacher interventions in lessons with demanding modelling tasks. In: M. Bosch (ed.), *CERME4 – Proceedings of the fourth conference of the European Society for Research in Mathematics Education.* Barcelona: Universitat Ramon Llull. pp.1623-1633.

Bowie, L.; Cronje, E.; Heany, F.; Maritz, P.; Olivier, G.; Rossouw, B. & Willers, S. 2012. *Platinum mathematics: Learner's book, 7.* Cape Town: Maskew Miller Longman.

Caron, F. & Belair, J. 2007. Exploring university students' competence in modelling. In: C. Haines, P. Galbraith, W. Blum & S. Khan, *Mathematical modelling (ICTMA 12): Education, engineering and economics.* Chichester, UK: Horwood Publishing. pp.120-129.

Corbin, J. & Strauss, A. 2008. Basics of qualitative research: Techniques and procedures for developing grounded theory (3rd edition). Thousand Oaks, CA: Sage. https://doi.org/10.4135/9781452230153

Davis, P.W. 1991. *Some views of mathematics in industry from focus groups.* SIAM Mathematics in Industry Project, Report 1. Philadelphia, PA: Society for Industrial and Applied Mathematics.

Davis, P. & Hersh, R. 1988. *Descartes dream: The world according to mathematics.* Penguin Books: London.

DBE (Department of Basic Education). 2011. *National curriculum statement (NCS): Curriculum and assessment policy statement (CAPS), further education and training phase Grades 10-12.* Pretoria: DBE.

DBE (Department of Basic Education). 2015. *National Senior Certificate: Mathematics P1, November 2015.* Pretoria: DBE.

De Oliveira, A.M.P. & Barbosa, J.C. 2010. Mathematical modelling and the teachers' tensions. In: R. Lesh, P. Galbraith, C.R. Haines & A. Hurford (eds.), *Modeling students' mathematical modeling competencies.* New York: Springer. pp.511-517. https://doi.org/10.1007/978-1-4419-0561-1_44

Galbraith, P. & Stillman, G. 2006. A framework for identifying student blockages during transitions during in the modelling process. *ZDM,* 38(2):143-162. https://doi.org/10.1007/BF02655886

Galbraith, P.; Stillman, G.; Brown, J. & Edwards, I. 2007. Facilitating middle secondary modelling competencies. In: C. Haines, P. Galbraith, W. Blum & S. Khan, *Mathematical modelling (ICTMA 12): Education, engineering and economics.* Chichester, UK: Horwood Publishing. pp.130-140. https://doi.org/10.1533/9780857099419.3.130

Giordano, F.R. & Weir, M.D. 1985. *A first course in mathematical modelling.* Monterey, CA: Brooks/Cole Publishing Company.

Haines, C.; Galbraith, P.; Blum, W. & Khan, S. (eds.). 2007 *Mathematical modelling (ICTMA 12): Education, engineering and economics.* Chichester: Horwood Publishing.

IEB (Independent Examination Board). 2014. *National senior certificate: Mathematics: Paper I.* Pretoria: IEB.

Julie, C. 2002. Making relevance relevant in mathematics teacher education. In: I. Vakalis, D. Hughes-Hallett, C. Kourouniotis, D. Quinney & C. Tzanakis (eds.), *Proceedings of the 2nd International Conference on the Teaching of Mathematics at the Undergraduate level,* University of Crete, Hersonissos, Crete, Greece, 1-6 July. CD-ROM published by Wiley Publishers.

Julie, C. 2004. Can the Ideal of the development of democratic competence be realized within realistic mathematics education? The case of South Africa. *The Mathematics Educator,* 14(2):34-37.

Julie, C. & Mudaly, V. 2007. Mathematical modelling of social issues in school mathematics in South Africa. In: W. Blum, P. Galbraith, H-W. Henn & M. Niss (eds.), *Modelling and applications in mathematics education: The 14th ICMI study.* New York: Springer. pp.503-510. https://doi.org/10.1007/978-0-387-29822-1_58

Julie, C. 2015. Learners' dealing with a financial applications-like problem in a high-stakes school-leaving mathematics examination. In: G.A. Stillman, W. Blum & M.S. Biembengut (eds.), *Mathematical modelling in education research and practice: Cultural, social and cognitive influences.* Dordrecht: Springer. pp.477-486. https://doi.org/10.1007/978-3-319-18272-8_40

Maaß, K. 2006. What are modelling competencies? *Mathematics Education Review (ZDM),* 38(2):113-142. https://doi.org/10.1007/BF02655885

Ng, K.E.D. 2013. Teacher readiness in mathematical modelling: Are there differences between pre-service and in-service teachers? In: G.A. Stillman, G. Kaiser, W. Blum, & J.P. Brown (eds.), *Teaching mathematical modelling: Connecting research and practice.* New York: Springer. pp.339-348. https://doi.org/10.1007/978-94-007-6540-5_28

Niss, M. 2010. Modeling a crucial aspect of students' mathematical modeling. In: R. Lesh, P. Galbraith, C.R. Haines & A. Hurford (eds.), *Modeling students' mathematical modeling competencies.* New York: Springer. pp.43-59.

Niss, M.; Blum, W. & Galbraith, P. 2007. Introduction. In: W. Blum, P. Galbraith, H-W. Henn & M. Niss (eds.), *Modelling and applications in mathematics education: The 14th ICMI study.* New York: Springer. pp.5-32. https://doi.org/10.1007/978-1-4419-0561-1_4

Olivier, A. (n.d.). Activity scarce metal. Notes handed out at teacher professional development meeting for the Western Cape Education Department.

OECD (Organisation for Economic Co-operation and Development). 2002. *Definition and selection of competences* (DeSeCo): *Theoretical and conceptual foundations: strategy paper.* Neuchatal, Switzerland: Swiss Federal Statistical Office. Availabe: http://bit.ly/2MM5Rtw (Accessed 29 July 2019).

Sen, A. 1976. Poverty: An ordinal approach to measurement. *Econometrica*, 44(2):219-232. https://doi.org/10.2307/1912718

Stillman, G. 1998. The emperor's new clothes? Teaching and assessment of mathematical applications at the senior secondary level. In: P. Galbraith, W. Blum, G. Booker & I.D. Huntley (eds.), *Mathematical modelling: Teaching and assessment in a technology-rich world.* Chichester, UK: Horwood. pp.243-253.

Stillman, Brown & Galbraith. 2010. Identifying challenges within transition phases of mathematical modeling activities at year 9. In: R. Lesh, P.L. Galbraith, C.R. Haines & A. Hurford (eds.), *Modeling students' mathematical modeling competencies.* New York: Springer. pp.385-296. https://doi.org/10.1007/978-1-4419-0561-1_33

Thompson, T. 2015. From big bang to big rip: How will the universe end? *New Statesman,* 2 July 2015. Available: http://bit.ly/2ZA21py (Accessed 30 July 2019).

| 07 |

PROFESSIONAL LEARNING IN THIRD SPACES

Raymond Smith

Introduction

This chapter reflects on a professional development project based on an alliance between the UWC, selected schools in its surrounding communities and some mathematics curriculum advisors. The chapter refers to the metaphor of a third space, which represents the intersection of the discourse spaces of academics in mathematics education, mathematics specialists and teachers. This metaphor provides the analytical tools to reflect on the experiences and learnings within the LEDIMTALI community of teacher educators, departmental officials and teachers. This chapter aims to contribute to the literature on effective teaching development and the search for teacher development models that may be effectively implemented across the wider educational landscape.

Education administrators and teacher educators need tools to help them think about teacher learning, to design activities and programmes that foster it, and to assess the results of their work with pre-service and in-service teachers (Levine, 2010). This chapter argues that to develop these tools, we need to blur the boundaries between theory and practice and create a common discourse space for the expertise to be shared between academics and practitioners.

According to Lave (1996:8), the theory of learning situated in a community of practice can be supported by four knowledge and learning premises:

1. Knowledge always undergoes construction and transformation in use.
2. Learning is an integral aspect of activity in and with the world at all times. That learning occurs is not problematic.
3. What is learned is always complexly problematic.
4. Acquisition of knowledge is not a simple matter of taking in knowledge; rather, things assumed to be natural categories, such as 'bodies of knowledge', 'learners', and 'cultural transmission', require re-conceptualisation as cultural, social products.

The mechanism of learning in discourse communities, such as third spaces, is 'systematic intentional inquiry' into all of the decisions, dilemmas, and kinds of knowledge that comprise the act of teaching (Levine, 2010:113). The way in which professional learning is afforded in this discourse community may include joint research projects, protocol-guided discussions between academics and practitioners, and joint consultations with groups of teachers. In this way, participants in third spaces become agentive constructors of shared knowledge as they jointly create or revise theoretical constructs that guide their work (Dunn, DePalma, Kinslow & Burger, 2013). Furthermore, this discourse community can engage with the prevailing discourse in schools to assist teachers to identify elements of their practice that are unexamined and portions of their professional knowledge that need upskilling and reskilling.

What are 'third spaces'?

This section explores a number of important features constitutive of a 'third space'. A third space represents the intersection of various discourse spaces (illustrated in Figure 7.1). In the context of CPD, a third space straddles 'academic space' and 'professional space' and leads to the construction of 'a professional learning space'. It is a collaborative space seeking to dismantle pre-conceived spaces for academics (didacticians), departmental officials (subject advisors) and practitioners (teachers) to address the conundrum of the separation of theory and practice.

The participants in the third space come from different discourse communities. According to Whitchurch (2008), and as Figure 7.1 illustrates, we may visualise this as the intersection of two discourse spaces.

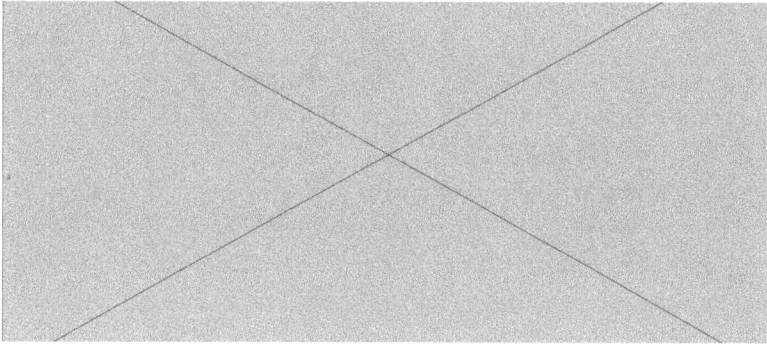

FIGURE 7.1   A third space as the intersection of various discourse spaces

Discourse communities have different concerns and motives.

Discourse space 1 represents the teaching fraternity. One of the motives in this discourse space is to improve the learning outcomes for all students. Thus the prevailing discourse in this community seeks solutions that afford immediacy of impact and pragmatism of implementation.

Discourse space 2 represents the academic research domain. Researchers in this domain might be interested in the appropriateness of the theories they are working with and being offered opportunities to trial them to advance or refute these theories or generate new hypotheses. Ur (1992) reflected on the different concerns and foci of academics and professionals. Table 7.1 tabulates these concerns.

TABLE 7.1   Academic versus professional

The academic	The professional
Is primarily concerned with abstract thought	Is primarily concerned with real-time action
Acts to refine thinking	Thinks to improve action
Is interested in finding out the truth	Is interested in finding out what works
Is not an immediate agent of real-world change	Is an immediate agent of real-world change
Is evaluated by publications (in the short term) and influence on real-world thought and action (in the long term)	Is evaluated by the extent to which change seen as valuable is brought about by action

Source: Adapted from Ur (1992:59)

An important feature of a third space is that it may be perceived as an interruptive and interrogative space, but at the same time, it represents a generative and innovative space (Benson, 2010; Stein & Coburn, 2005). It is interruptive in the sense that it requires participants to suspend their current views and works towards a co-constructed viewpoint. In addition, it requires the interrogation of various assumptions and arguments to be able to move towards a more consensual position. A well-constructed third space has the potential to accommodate a range of different activities in one virtual location offering a variety of working options, thus helping to enhance productivity and creativity. In this way, a third space facilitates a mode of articulation, a way of defining a productive and not merely a reflective disposition that may engender new possibilities (Dreaver, 2008; Stein & Coburn, 2005).

In summary, a third space creates an intersecting discourse community where various communities of practice could engage democratically, and in the words of Taylor and Klein, "disrupt the traditional power relationships and participate in a space where the roles of the university, school, teacher candidate and community could be reimagined" (Taylor & Klein, 2015:11). In this way, we experience the mobilisation of the different knowledges, expertise and skills which the various participants have to offer. Taylor and Klein provide the following useful insight:

> We realised for this to work, we would need to allow the third space to be dynamic, ambiguous, ever-shifting and always under construction. That required continuous generative conversations among all participants to share their knowledge, experiences and expertise that would lead to the co-construction of a blueprint for the programme. (Taylor & Klein, 2015:11)

The third space model of professional learning

The third space model of professional learning as proposed in this chapter draws extensively from research studies on the professional development of teachers. This is particularly apposite with regards to effective CPD criteria. In their review of the research on professional development programmes, Hawley and Valli (1999) identified eight principles for effective professional development. They posit that effective CPD programmes are driven by attention to goals and student performance and built upon teacher involvement in identifying learning needs and shaping the learning opportunities and processes. Furthermore, effective CPD should ideally be school-based emphasising job-embedded learning. Despite it being advised that collaborative inquiry into teaching and learning forms the key strategy to professional learning, CPD should be data-driven and theoretically grounded. In this way, both practitioner inputs and well as academic inputs form the backbone of professional learning in third spaces.

Noting the principles espoused here, the implications for professional learning in third spaces is quite clear: Professional learning as the object of CPD must be goal orientated and focused on knowledge construction activities that are research-based, data-informed and that contribute to meeting individual, local as well as the broader organisational needs.

The following section highlights the professional identities of the participants in a third space and the progressive attributes of the various participants.

Professional identity in third spaces

Whitchurch (2008) identified four types of professionals and presents a typology, as indicated in Table 7.2.

TABLE 7.2   A typology of professional identities in the CPD domain

Categories of identity	Characteristics	Examples in the context of CPD
Bounded professionals	Work within clear structural boundaries	Teachers, lecturers
Cross-boundary professionals	Actively use boundaries for strategic advantage and institutional capacity building	Subject advisors, CPD providers/facilitators
Unbounded professionals	Disregard boundaries to focus on broadly-based projects and institutional development	Project managers, institutional managers, professionals within funding agencies
Blended professionals	Dedicated appointments spanning professional and academic domains	Researchers; research chairs; project workers

Source Adapted from Whitchurch (2008:384)

The value of the typology as shown in Table 7.2 lies in the language it provides to describe roles and responsibilities in the third space created for CPD purposes. It also serves as an analytic lens for exploring the qualities needed in brokers who are the agents facilitating professional learning activities in the third space. Participants in a third space may be exposed to new ideas and practices that affect the meaning members create as they negotiate new learning within their communities. This process happens through three mechanisms, namely:

ø   boundary practices, which are regular ongoing interactions across boundaries and within joint activities;

ø   brokers, who are individuals holding membership in multiple communities and can carry practice between them; and

Ø boundary objects, which are artefacts, terms, concepts, or documents that travel across the boundaries of one community into another.

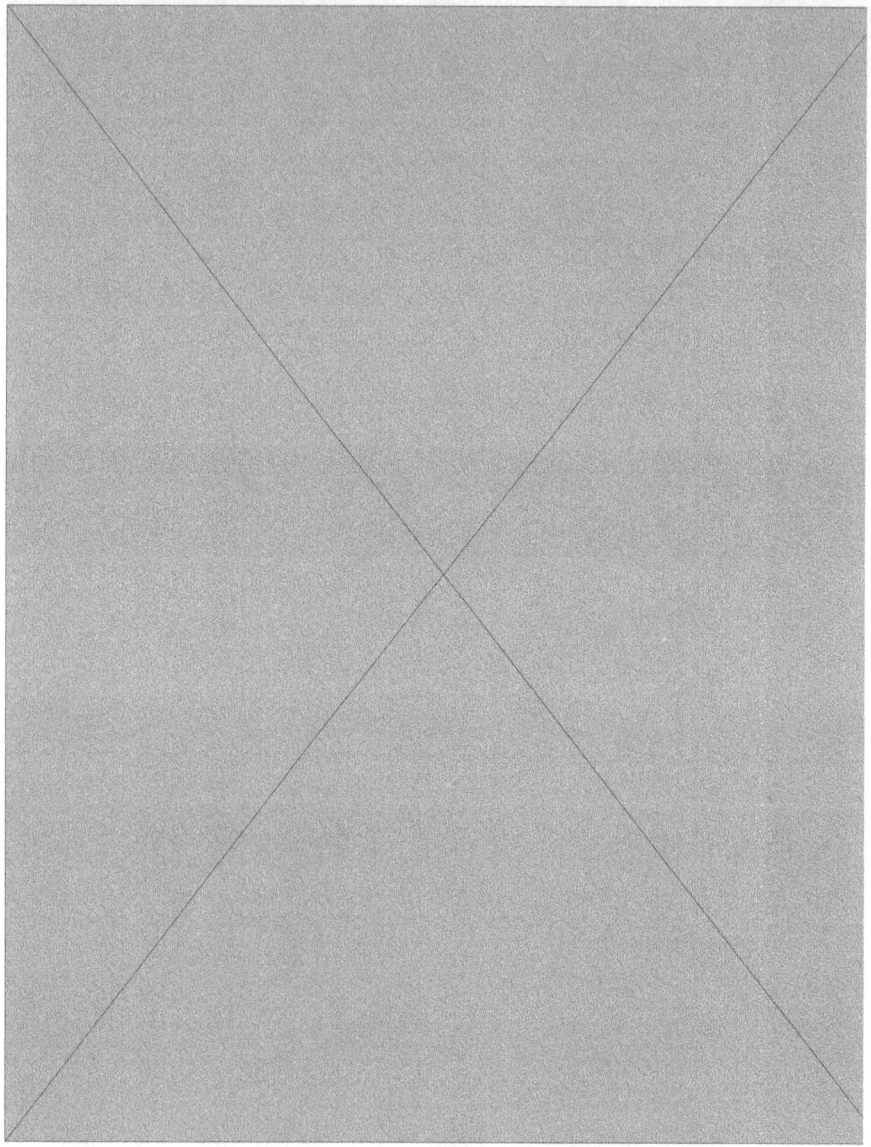

FIGURE 7.2   The architecture of a third space

Source   Adapted from Stein and Coburn (2005:15)

As illustrated in Figure 7.2, there is acknowledgement and appreciation of the knowledge and theories that both teachers and didacticians bring from their respective discourse communities and this contributes to the shared knowledge and theories held in the third space. Together, teachers and didacticians apply and test their knowledge and theories when they use them to investigate authentic problems of practice in the third space. The consequence of this inquiry is new learning for members of both communities of practice. This learning drives the decisions each makes about future action. The long-term consequence should be improved knowledge and practice in each discourse community, resulting in collaborative learning benefits, such as better outcomes for students and increased student retention, and creates a stronger social support system for teachers. Eventually, it is a process through which all participants can work towards the development of a common theory of improvement.

Stein and Coburn (2005) identify several criteria about the conditions required to foster learning in the third space:

Ø It is essential to choose boundary practices that create pathways for ideas, approaches and artefacts to flow into the development process and for new learning to flow back to the home communities.

Ø It is important to address issues about status and authority in order to develop productive working relationships across institutions.

Ø The boundary objects need to be targeted enough to keep the work focused, yet flexible enough to allow for the negotiation of meaning between individuals from different communities.

Ø If the goal is for practitioners in similar communities of practice to move towards roughly similar forms of practice, they also need opportunities to negotiate meaning with those who develop boundary objects. For example, it is not enough for developers to disseminate boundary objects, such as new curriculum documents, to multiple school communities and expect uniform understanding and use of these boundary objects.

The message conveyed to the different discourse spaces coming together in the context of a third space is that there are certain hurdles to be negotiated in order to create the conditions for meaningful dialogue and sense-making. It is thus beholden on all stakeholders concerned with CPD to take cognisance of the above requirements for a progressive and generative culture within the third space to emerge.

Boundaries and boundary objects

Collaboration and inter-disciplinary innovation is a central feature of the post-modern society. Hence, this chapter proposes the concept of boundary-crossing as one way of tackling problems in the education sector. According to Bowker and

Star (1999), the creation and management of boundary objects is a key process in developing and maintaining a coherent discourse across different communities of knowledge practice. For this reason, it is pertinent to consider the notion of boundary objects. A boundary object is a sociological construct which signifies representational forms – material or conceptual – that can be shared between different communities (Star & Griesemer, 1989). The importance of boundary objects for this research is that they afford communication, knowledge sharing and knowledge generation across the boundaries of the different communities of knowledge practice in the educational sector. In so doing, boundary objects facilitate the adoption and implementation of solution strategies and innovation in education. Fox (2011:86) posits that "an effective boundary object might even succeed in bringing harmony to dissensus, or peace to a conflicted situation."

What are boundary objects?

As mentioned, boundary objects are artefacts or abstract constructs that facilitate knowledge sharing and knowledge generation between different communities of knowledge practice, for example, between teachers and students, subject advisors and teachers, and the education department and academic institutions. In the context of this chapter, boundary objects can be representations, documents, technologies, abstractions or metaphors that have the power to 'speak' to different role-players in education. Dirkink-Holmfeld (2006) argues that boundary objects can potentially facilitate and improve innovation across various disciplinary fields. This potential resides in the fact that boundary objects facilitate shared meaning-making.

Facilitative and inhibitory boundary objects

According to Carlisle (2002), effective boundary objects provide a shared language, allow concerns to be expressed, and afford growth in knowledge.

However, Fox (2011) maintains that some boundary objects may act negatively by inhibiting the adoption of a particular innovative strategy. Some examples that come to mind are inflexible policies, vague strategic plans or goals, and abstract theories.

A typology of boundary objects

It must be clearly understood that intrinsically, boundary objects do not by themselves make a difference or affect change in the participants in a third space. For effective implementation, participants need to understand what these boundary objects represent and how they can be effectively utilised to enhance professional learning. In this regard, I find that the categorisation of boundary objects developed by Star and Griesemer (1989) contributes to an efficacious understanding of the attributes of boundary objects that may lead to effective deployment. The scheme proposed by Star and Griesemer (1989) suggest a fourfold categorisation of boundary objects, namely:

Ø knowledge repositories;

Ø ideal types;

Ø common objects; and

Ø standardised templates.

Table 7.3 gives a succinct summary of these categories of boundary objects. Furthermore, the table provides some suggestions as to how they may be effectively deployed in actively engaging in collaborative sense-making within a third space. The table also provides concrete examples of useful boundary objects within the third space discourse community.

TABLE 7.3   A typology of boundary objects

Category of boundary objects	Deployment purpose	Examples of boundary objects
Knowledge repositories	Provide access for different role-players to each other's knowledge sources and resources	Reports and research data
		Institutional memories
		Skillsets
		Diagnostic reports from different assessments
Ideal types	Provide a broad blueprint for different positions and possibilities	Strategic plans
		Curriculum maps, concept maps, pacesetters and lesson plans
		Common and exemplar assessment tasks
Common objects	Reside in all the sectors but may contain different content	Contextual information, standard operating procedures
		Organisational maps
Standardised templates	Provide a means for all participants to obtain evidence and communicate concerns and proposals	Questionnaires
		Project plans
		Problem statements
		Funding proposals

Source Adapted from Star and Griesemer (1989:410)

The effective deployment of the boundary objects shown in the table requires skilful persons with a deep understanding of the process of deploying them. This process is referred to as brokering.

Brokering

Etienne Wenger describes brokering as follows:

> The job of brokering is complex. It involves processes of translation, coordination and alignment between perspectives. It requires enough legitimacy to influence the development of a practice ... it also requires the ability to link practices by facilitating transactions between them and to cause learning by introducing into a practice, elements of another. (Wenger, 1998:109)

Brokers are participants in a third space with access to the different communities of knowledge practice who can facilitate effective connections between the different groupings (Kimble et al., 2010). The key function of a broker is to manage collaboration and knowledge sharing. To be effective as a broker, the individual or individuals need to have authority within all the groups and possess the capacity to manage inter-group relations and trust. Kimble et al. (2010) posit that over time the broker's activities may lead to the development of a repertoire of shared resources, such as rules and procedures as well as the boundary objects used by the group.

Having described the architecture of a third space and the key actors in facilitating its activities, we now consider examples of third spaces in the context of CPD.

Examples of third spaces

The first noteworthy example of third spaces in the context of CPD is that of PLCs as they are commonly denoted. PLCs have become the model of choice for CPD in many countries, in particular within the South African context. The second example of how a third space may be enacted is that of the LEDIMTALI project.

Professional learning communities as an example of a third space

In addition to the formation of partnerships, third spaces are characterised by the establishment of PLCs. Koellner-Clark and Borko maintain that establishing PLCs is equivalent to:

> creating a safe environment for teachers to grapple with difficult content and pedagogical issues, developing sustained relationships among teachers in the community, encouraging participants to listen carefully to each other's ideas and perspectives, equally distributing social and intellectual work within the community, and fostering a commitment to helping others within the group learn

and develop both intellectually and in their teaching practice. (Koellner-Clark & Borko, 2004:223)

The phrase 'professional learning community' consists of three words, each encompassing an important meaning (Stoll and Louis, 2005). These meanings are understood as indicating the importance, and the power of a PLC as a supportive mechanism for continuous professional teacher development (Burke, 2013). In this chapter we may conceive these meanings as follows:

Ø 'Professional' implies that the community's work is underpinned by a specialised knowledge base which is discipline-specific. Furthermore, the term 'professional' indicates that the work of teachers requires special education and training in order to practice, is service-oriented and governed by a strong identity of professional commitment.

Ø 'Learning' connotes an emphasis on improvement. Professional learning is focussed on improving knowledge and skills, which will enhance the quality of teaching, and as a consequence, the learning that takes place in classrooms.

Ø 'Community' emphasises the collaborative, supportive and sharing engagement between members. It points towards a relational trust between members, which allows them to share personal practice.

In summary, Jennifer Hefner expresses the focus on PLCs eloquently:

> By using the term professional learning community we signify our interest not only in discrete acts of teacher sharing, but in the establishment of a school-wide culture that makes collaboration expected, inclusive, genuine, ongoing, and focused on critically examining practice to improve student outcomes. (Hefner, 2011:24)

LEDIMTALI as a prototype of a third space in the South African context

The LEDIMTALI project is a project concerned with, amongst others, the improvement of achievement in mathematics and increasing the number of students writing mathematics as an examination subject in the NSC examination.

LEDIMTALI is a collaborative project at the University of the Western Cape between schools and the university. The membership of the collaborative partnership constituted of:

Ø mathematics teachers from ten schools;

Ø teacher educators from the UWC Faculty of Education;

Ø mathematics educators from the UCT, Stellenbosch University, and the Cape Peninsula University of Technology (CPUT);

Ø mathematicians from the Department of Mathematics at UWC and the Faculty of Engineering at CPUT; and

Ø officials from the WCED.

It is noteworthy that two other educational districts (one in another province of the country), show interest in the support that this project has to offer. This shows that the project may have certain alluring features that both practitioners and bureaucratic managers find appealing and promising.

The project collaborates with mathematics teachers working in schools in socio-economically deprived environments. These areas are characterised by problems such as high rates of unemployment, gangsterism, drug abuse, single-parent households and high rates of teenage pregnancies.

LEDIMTALI focuses on the development of teaching for the improvement of achievement results in mathematics. 'Local evidence-driven' signifies that the development of teaching flows from the 'local' concerns, dilemmas and practices of the participating teachers. This does not mean that the participating academics do not offer and overlay their knowledge and experiences as possibilities for consideration for addressing concerns and dilemmas and enhancing teaching practices. In broad terms, LEDIMTALI represents a third space.

The major epistemological underpinnings of the project are twofold:

1. The legitimate school mathematics knowledge is the knowledge that is assessed in time-restricted high-stakes school examinations.

2. Teaching should focus on thoughtful emphasis on the "practicing and consolidation of concepts and procedures" and 'mathematical process skills' driven by the weighting accorded to these two components in the time-restricted high-stakes examinations (Okitowamba, Julie & Mbekwa, 2018).

The aspect of 'legitimate mathematical knowledge' as espoused in this paragraph implies that the local South African syllabus as prescribed by the DBE is taken as the basis for teaching, learning and assessment and hence this should inform CPTD interventions. The practical implication of this is that the current school syllabus for Grades 10-12, as well as the approved textbooks and exemplar examination papers, past NSC Mathematics examination papers and diagnostic reports of these examinations are important boundary objects, which sets the standards for PD inputs regarding specific subject content knowledge as well as pedagogical content knowledge of teachers. These boundary objects also inform specific strategies employed in the project such as designing activities for spiral revision, productive practicing as well as assessments for learning.

The LEDIMTALI project adopted an epistemological stance, which asserts that the legitimate mathematical knowledge for schools is the mathematics embedded in the South African curriculum and which is examined in the NSC examination. This then forms the basis of the content and pedagogical content knowledge on which all professional development (PD) activities are based. The rationale behind this stance is that it foregrounds the knowledge that all students will have to demonstrate during their final NSC examination at the end of Grade 12.

This epistemological stance is supported by particular epistemic commitments, such as intentional teaching, productive practices, spiral revision and assessment for learning and working with feedback.

The varying goals of the different role-players necessitating the importance of partnerships cannot be underestimated. Partnerships have great value as they break down the isolation teachers experience with the broader community of educators (Day, 1999; Zarkovic & Rutar, 2006). For the community of educators, it is not unusual for partnerships between university academics and teachers to arise. This partnership model assumes that each party has something to contribute to the professional learning of the other.

Such partnerships are usually initiated by university academics for the purpose of professional development and educational research. This is because university staff are in a much more favourable position of facilitating such partnerships. The danger exists that an unequal power relationship may exist between academics and practitioners and that academic knowledge is deemed to be privileged over practitioner knowledge. It is at this level that the notion of 'third spaces' becomes powerful in levelling the playing field. This is when both parties are positioned to interact as equals and recognise each other's expertise. At this stage, one can talk about a two-way model of reciprocity. This is what Zarkovic and Rutar suggest when they state the following:

> If we want to take a step further, more power should be given to teachers. It is also important that teachers start with problems, which have personal meaning and importance for them. In such a way they become more engaged than while solving the problems that are alienated from them and imposed by external institutions. (Zarkovic & Rutar, 2006:130)

Different types of partnerships have been posited and Zarkovic and Rutar offer a quartic typology. These four types are collaborative, complementary, developmental, and implementational partnerships. Zarkovic and Rutar

(2006) distinguish between collaborative and complementary partnerships. In complementary partnership, both partners have separate and complementary responsibilities but there is no attempt to bring the two in dialogue. A collaborative partnership, on the other hand, is seen as a common endeavour by both parties who possess a different body of knowledge. Outcomes of such partnerships and collaboration might be negotiated expectations, collaborative planning, sharing of expertise, diversity of perspectives and viewpoints, knowledge generation and development of trust.

Day describes two other types of partnerships: developmental partnership (responsive and evolutionary), and implementational partnership (imposed, formal, mechanistic and with a specific, brief and limited lifespan for action). Developmental partnerships will often begin with the cultures of 'contrived collegiality' typical of implementational partnership, but they have a greater learning potential because the ownership of the theme and the process is controlled by participants themselves (Day 1999, cited in Zarkovic & Rutar, 2006).

Whilst we may typify LEDIMTALI as a collaborative partnership because of the formal agreements between participating institutions, it also exhibits much of the attributes of a developmental partnership. It is in this attribute from a developmental partnership that we locate LEDIMTALI's connectivity to the 'third space' conceptualisation as illustrated in Figure 7.1 described earlier in this chapter. The rationale behind this connectivity to third spaces is founded on the principle of shared ownership of processes and emerging themes. This happens without participants sacrificing their professional identities and thus engaging as equals in constructing shared professional learning.

The LEDIMTALI project combines both these examples to create a model of continuing professional development. We will now discuss some of the learnings from the LEDIMTALI project.

Learnings from the LEDIMTALI project as a third space

Teachers in the project meet about nine times a year for afternoon workshops every month during the first three school terms. Participants also get together three times a year over a weekend in a teacher institute. A teacher institute in the context of this LEDIMTALI is a weekend breakaway workshop, starting on a Friday afternoon and ending on a Sunday afternoon. It normally takes place in a suitable conference centre. The teacher institutes provide space where the members of the collaborative partnership can interact without the distractions of time, family commitments and other extra-mural responsibilities. In this context, they are immersed in their engagement with professional learning activities, relationship building and educational problem-solving.

The paragraphs that follow consider some pertinent aspects emanating from the workshops.

Professional learning

In the spirit of the third space notion, it is not only teachers who gain from the partnership. In a recent mid-term evaluation of the project, didacticians also commented on their learning. To date, their major learning is linked to their practice. One didactician, for example, commented that:

> One didactician   [T]he experience that we have, with those groups of students and those groups of teachers, impacts us because we are educators of pre-service teachers. We change and reconstruct where we put things there, in our curriculum.

Pedagogical change and innovation may be espoused as two important outputs of in-service professional learning programmes at any learning and teaching institution. Proponents of practice-based professional learning argue that learning experiences that are highly connected to and contextualised in professional practice can better enable mathematics teachers and didacticians to make the kinds of complex, nuanced judgements required in teaching (Goos, Dole & Makar, 2007; Silver, Clark, Ghousseini, Charalambous & Sealy, 2007).

Apart from the structured professional learning activities, LEDIMTALI provided informal learning opportunities, especially during workshops and at institutes, by creating an increased awareness in teachers about pedagogy. Teachers commented that even outside of the formal activities, they still found opportunities to learn professionally. One teacher articulated this as follows:

> Teacher 1   We believe the group has achieved some sense of trust due to the existence of a conducive environment where openness to share ideas on teaching math, there is voluntary participation irrespective of diverging groups. Group discussions are collegial, open and different opinions are respected. Members of the groups interact freely within the workshops and outside the workshops.

Teachers are encouraged to be reflective and critical of their practice by engaging each other during break times at teacher institutes. The interview quoted above ascribes this as a consequence of the trusting relationships that developed. Another teacher commented:

Teacher 3   We feel free to share ideas. One can only do that if you trust people. We feel free to differ on opinion whether it's right or wrong. This implies a two-way relationship of trust. We can be ourselves when we work and interact with each other – openness to be who we are.

Moreover, teachers are acknowledged and their successes in implementing new skills celebrated. Acknowledging efforts and celebrating successes encourages other teachers to implement similar initiatives. These are generally reported on when teachers meet again during the next institute. LEDIMTALI also strives to distribute leadership across teachers who show leadership qualities, for instance, in implementing peer observation, leading grade meetings and planning sessions.

Collaboration as the backbone of a third space

Collaboration in the context of third spaces may be understood as a voluntary, mutually beneficial engagement between educators from an HEI and schools, in other words, educators from various institutions, sharing knowledge and resources to solve educational problems. The concepts of relationships, capacity, commitment and resources frame our understanding of what collaboration within a third space might require.

Collaboration may effectively lead to relationship building among participants and so enhance the efficacy and sustainability of the collaborative partnership. At the same time, sustainable collaboration depends on the type of social and professional relationships that emerges in this process. Concomitantly, collaboration supports personal as well as collective capacity development within the PLC members.

Finally, we contend that collaboration affords available resources and expertise amongst the participants, leading to innovation and context-specific solutions to educational problems faced by members of the PLC. Goddard, Goddard and Tschannen-Moran advance the following thesis: "When teachers collaborate, they share experiences and knowledge that can promote learning for instructional improvement … such learning can help teachers solve educational problems, which in turn has the potential to benefit students academically" (Goddard et al., 2007:891).

Instances of this statement were also found in this collaborative partnership. Teachers in the project observed:

Teacher 1   All members of the group contribute in discussions and all members contribute when exam papers are set.

Teacher 2   The sharing of ideas, learning from other people's experiences as well as sharing resources are some important areas of collaboration among us.

Teacher 3   We work together during group activities at the workshops to formulate grade lesson plans, common tasks and examinations.

Thus, in LEDIMTALI, collaboration became a vehicle for sharing ideas and resources as well as providing mutual support. Collaboration during workshops or teacher institutes usually took the form of group work – sometimes across different grades and sometimes within grades. Within grades, activities centred around lesson planning, constructing common assessments and refining pedagogical approaches, such as intentional teaching and spiral revision. One of the aims was to provide materials and classroom strategies that can effectively support teachers as they develop their pedagogical practice.

However, it was not always without its challenges. Some teachers also felt intimidated by the perceived 'superior knowledge' of others:

Teacher 1   Challenges were in the perceptions or impressions about the greater expertise/ knowledge of the professors/ lecturers / curriculum advisors leading to teachers feeling intimidated during group activities/looking at these people for leadership and initiative. Some people feel inferior whilst others think they are superior – people are intimidated by this.

Trust as the "glue that binds us in a third space"

The importance of working together in a trusting environment assisted us in LEDIMTALI to overcome the challenges mentioned in the previous extract. There are strong signals the concept of a third space is taking root in the LEDIMTALI collaborative partnership:

Teacher 4   We feel free to share ideas. One can only do that if you trust people. We feel free to differ on opinion whether it's right or wrong. This implies a two-way relationship of trust.

Teacher 5   We can be ourselves when we work and interact with each other – openness to be who we are.

This observation is confirmed by the sentiment expressed by a group of teachers interviewed for this chapter.

LEDIMTALI   A safe space has been created for teachers to share and
focus group 1   learn from each other even sharing our uncertainties and challenges. We believe that LEDIMTALI teachers have achieved some sense of trust due to the existence of a conducive environment where openness to share ideas on teaching mathematics, there is voluntary participation irrespective of the diversity in the composition of the groups. Although different opinions exist every-one agreed that the purpose is to improve teaching and learning. So it makes it easy to cope with divergent views. This contributes to a sense of trust between participants. Group discussions are collegial, open and different opinions are respected. Members of the groups interact freely within the workshops and outside the workshops. We feel free to share ideas. One can only do that if you trust people. We feel free to differ on opinion whether it's right or wrong. This implies a two-way relationship of trust.

Developing a common language to discuss practice

"Teachers learn by doing, reading, and reflecting (just as students do); by collaborating with other teachers; by looking closely at students and their work; and by sharing what they see" (Darling-Hammond & McLaughlin, 2011:83). It is in the "sharing what they see" where it becomes crucially important to develop a common language to discuss practice. This happened organically in LEDIMTALI where issues, such as "intentional teaching, spiral revision, and assessment for learning" became part of our common vocabulary.

Shared ownership of processes and products and boundary objects

We concur with Hargreaves and Giles (2003:9) that a PLC [or third space community] brings together the knowledge, skills and dispositions of university lecturers, subject advisors and teachers in a partnership that has the potential to promote shared learning and improvement. They maintain that a strong professional learning community is a social process for turning information into knowledge.

Collaboration and inter-disciplinary innovation is a central feature of the post-modern society. According to Bowker and Star (1999), the creation and management of boundary objects is a key process in developing and maintaining a coherent discourse across different communities of knowledge practice. This knowledge and the concomitant artefacts produced become the shared property of the learning community.

In the LEDIMTALI project, all participants engaging in knowledge and artefact production facilitated this shared ownership. In fact, the concepts of 'assessment for learning', 'learning intentions', 'success criteria', etc. was introduced by a practicing teacher who is a head of department at one of the schools. These concepts became embedded in our discourse and guided the development of lesson frameworks. Similarly, every participant owns the concept of spiral revision and the artefacts that emanated from this approach. All these artefacts became important boundary objects in the course of the professional learning and professional growth of all the participants.

Brokers are essential to ensure that third spaces are effective and efficient in terms of their mandate. The key function of a broker is to manage collaboration and knowledge sharing. Also, the broker must have the respect and trust of all groupings. They must be knowledgeable and strategically adaptive in order to manage inter-group relations.

Conclusion

This chapter highlighted some of the lessons learnt in the LEDIMTALI project as an example of a third space where teacher educators, departmental support staff and teachers meet for the purpose of professional learning and development. In reflecting on the road we have travelled together since 2014 in the LEDIMTALI project, the following stand out as important lessons for us and others engaged in continuous professional development:

Changing our perspective from professional development to professional learning.

1. Collaboration as the backbone of this partnership.
2. Trust as the glue that binds us into a third space.
3. Developing a shared language to discuss teaching practice.
4. Shared ownership of processes and products and boundary objects.
5. The selection of an appropriate broker.

The model of in-service professional development described in this chapter holds promise as an effective means for developing FET mathematics teachers, especially in the current context of underperformance in mathematics at some schools. At the same time, university lecturers preparing these teachers for practice benefit professionally and grow in their roles as didacticians and professional leaders in teacher education. The model also affords educational administrators to be co-developers, as well as co-students in the entire process.

References

Benson, S. 2010. I don't know if that'd be English or not: Third space theory and literacy instruction. *Journal of Adolescent & Adult Literacy*, 53(7):555-563. https://doi.org/10.1598/JAAL.53.7.3

Bowker, G.C. & Star, S.L. 1999. *Sorting things out: Classification and its consequences*. Cambridge, MA: MIT Press.

Burke, B.M. 2006. Theory meets practice: A case study of pre-service world language teachers in U.S. secondary schools. *Foreign Language Annals*, 39(1):148-166. https://doi.org/10.1111/j.1944-9720.2006.tb02255.x

Burke, B.M. 2013. Experiential professional development: A model for meaningful and long-lasting change in classrooms. *Journal of Experiential Education*, 36(3):247-263. https://doi.org/10.1177/1053825913489103

Carlisle, P.R. 2002. A pragmatic view of knowledge and boundaries: boundary objects in the development of new product development. *Organisational Science*, 13(4):442-455. whttps://doi.org/10.1287/orsc.13.4.442.2953

Darling-Hammond, L. & McLaughlin, M.W. 2011. Policies that support professional development in an era of reform. *Phi Delta Kappan*, 92(6):81-92. https://doi.org/10.1177/003172171109200622

Day, C. 1999. *Developing teachers, the challenges of lifelong learning*. London: Falmer Press.

Dirkink-Holmfeld, L. 2006. *Designing for collaboration and mutual negotiation of meaning – boundary objects in networked learning*. Paper presented at an international research conference on 'Relations in networks and networked learning', Lancaster University, 10-12 April 2006.

Dreaver, K. 2008. Joint inquiry in the third space. In: K. Dreaver & S. Chiaroni, *Improving teacher educator learning and practice*. Wellington, NZ: Ministry of Education. pp.169-171. Available: http://bit.ly/2zvLSXp (Accessed 29 July 2019).

Dunn, K.; DePalma, D.G.; Kinslow, G. & Burger, T.B. 2013. The critical role of field placement in teacher education: Working toward a model of success. *Impact on Instructional Improvement*, 38(1):33-38.

Fox, N.J. 2011. Boundary objects, social meanings and the success of new technologies. *Sociology*, 45(1):70-84. https://doi.org/10.1177/0038038510387196

Goddard, Y.; Goddard, R. & Tschannen-Moran, M. 2007. A theoretical and empirical investigation of teacher collaboration for school improvement and student achievement in public elementary schools. *The Teachers College Record*, 109(4):877-896.

Goos, M.; Dole, S. & Makar, K. 2007. Supporting an investigative approach to teaching secondary school mathematics: A professional development model. In: J. Watson & K. Beswick (eds.), *Mathematics: Essential research, essential practice* (Vol. 1), *Proceedings of the 30th annual conference of the Mathematics Education Research Group of Australasia.* Adelaide: MERGA. pp.325-334.

Hargreaves, A. & Giles, C. 2003. The knowledge society school: An endangered entity. In: A. Hargreaves (ed.), *Teaching in the knowledge society: Education in the age of insecurity.* Maidenhead and Philadelphia: Open University Press. pp.127-159.

Hawley, W. & Valli, L. 1999. The essentials of effective professional development: A new consensus. In: L. Darling-Hammond & G. Sykes (eds.), *Teaching as the learning profession: Handbook of policy and practice.* San Francisco: Jossey-Bass. pp.127-150.

Hefner, J.F. 2011. *A case study of a professional learning community: An investigation of sustainability within a rural elementary school.* Doctoral dissertation, Appalachian State University, Boone, NC.

Kimble, C.; Grenier, C. & Goglio-Primard, K. 2010. Innovation and knowledge sharing across Professional boundaries: Political interplay between boundary objects and brokers. *International Journal of Information Management*, 30(5):437-444. https://doi.org/10.1016/j.ijinfomgt.2010.02.002

Koellner-Clark, K. & Borko, H. 2004. Establishing a professional learning community among middle school mathematics teachers. In: M. Hoines & A.B. Fuglestad (eds.), *Proceedings of the 28th conference of the international group for the psychology of mathematics education.* Bergen, Norway: Bergen University College.

Lave, J. 1996. The practice of learning. In: S. Chaiklin & J. Lave (eds.), *Understanding practice: Perspectives on activity and context.* New York: Cambridge University Press. pp.3-32. https://doi.org/10.1017/CBO9780511625510.002

Levine, T. 2010. Tools for the study and design of collaborative teacher learning: The affordances of different conceptions of teacher community and activity theory. *Teacher Education Quarterly*, 37(1):109-130.

Okitowamba, O.; Julie, C. & Mbekwa, M. 2018. The effects of examination-driven teaching on mathematics achievement in Grade 10 school-based high-stakes examinations. *Pythagoras*, 39(1):1-10. https://doi.org/10.4102/pythagoras.v39i1.377

Silver, E.A.; Clark, L.M.; Ghousseini, H.N.; Charalambous, C.Y. & Sealy, J.T. 2007. Where is the mathematics? Examining teachers' mathematical learning opportunities in practice-based professional learning tasks. *Journal of Mathematics Teacher Education*, 10:261-277. https://doi.org/10.1007/s10857-007-9039-7

Star, S. & Griesemer, J. 1989. Institutional ecology, 'translations' and boundary objects: Amateurs and professionals in Berkeley's Museum of Vertebrate Zoology 1907-1939. *Social studies in Science*, 19(3):379-420. https://doi.org/10.1177/030631289019003001

Stein, M. & Coburn, C. 2005. Toward producing usable knowledge for the Improvement of educational practice: A conceptual framework. Paper presented at the annual meeting of the American Educational Research Association, Montreal, 11-15 April 2005.

Stoll, L. & Louis K.S. 2007. Professional learning communities: Elaborating new approaches. In: L. Stoll & K.S. Louis (eds.), *Professional learning communities: Divergence, depth and dilemmas*. Maidenhead: Open University Press & McGraw Hill Education.

Taylor, M. & Klein, E.J. 2015. *A year in the life of a third space urban teacher residency: Using inquiry to reinvent teacher education*. Rotterdam/Boston/Taipei: Sense Publishers. https://doi.org/10.1007/978-94-6300-253-0

Ur, P. 1992. Teacher learning. *ELT journal*, 46(1):56-61. https://doi.org/10.1093/elt/46.1.56

Wenger, E. 1998. *Communities of practice: Learning, meaning, and identity*. New York: Cambridge University Press. https://doi.org/10.1017/CBO9780511803932

Whitchurch, C. 2008. Shifting identities and blurring the boundaries: The emergence of third space professionals in UK higher education. *Higher Education quarterly*, 62(4):377-396. https://doi.org/10.1111/j.1468-2273.2008.00387.x

Zarkovic, B. & Rutar, R. 2006. Are we ready for a step further-learning partnership and learning communities of teachers? In: M. Brejc (ed.), *Co-operative partnerships in teacher education: Proceedings of the 31st annual ATEE conference*. Portorož, Slovenia: National School for Leadership in Education. pp.125-133.

| 08 |

PROFESSIONAL LEARNING COMMUNITIES, PROFESSIONAL LEARNING AND THE ROLE OF RELATIONAL AGENCY

Raymond Smith, Cyril Julie, Lorna Holtman & Juliana Smith

Introduction

PLCs are becoming the preferred model of CPTD across the world. The PLC model is predicated on the assumption that the collaborative capacity afforded by a community orientation enhances professional learning. McNichol (2013) states that reviews of the literature and some empirical investigations into different models of PD from around the world have formed a consensus in calling for more collaborative approaches to PD (Abu-Tineh & Sadiq, 2017; Eliahoo, 2017; Kennedy, 2006).

However, the literature does not adequately explain how community affordance underpins the learning process. In this chapter, we argue that there is a need for a better conceptualisation of the role of community in professional learning. Hence we suggest that the concept of relational agency may provide a useful mechanism to account for professional learning in a collaborative setting. This view is supported by the tenets of social learning (Lave & Wenger, 1991) that there is a relational interdependency, underpinning socio-cultural epistemologies.

Concern about the overuse of the word 'community'

Researchers, such as Grossman, Wineburg and Woolworth (2001:492), are concerned that the word 'community' has lost its meaning due to their observation that "community has become an obligatory appendage to every educational innovation." American educational researcher Richard DuFour similarly remarks that the concept of professional learning community is in vogue but worries that so many have leapt onto the bandwagon that the phrase now describes "every imaginable combination of individuals with an interest in education" (DuFour, 2004:6). DuFour further expresses the concern that the concept of community is "in danger of losing all meaning." Joel Westheimer found the literature on teacher community 'disappointingly vague', and warns that without richer and more careful conceptualisation, "the rhetoric of community is rendered ubiquitous and shallow" (Westheimer, 1999:3).

The LEDIMTALI PLC

The PLC that presented the testbed for this research resided in the LEDIMTALI project. This initiative brought teachers, teacher educators and subject advisors together to establish a PLC. The LEDIMTALI project was a partnership between UWC, the WCED and ten participating schools. Participation in this PLC provided both structure and opportunity for professional learning. By structure, we refer to both physical, financial and human resources required for enacting professional development activities. Opportunity for learning is the affordance for enabling teachers to access these resources.

Teachers were able to interact and collaborate within a common practice and share their practice and to learn from the practices of others. They experienced professional learning through internal expertise (as represented by the teachers) coupled with external expertise (as represented by the departmental support officials as well as the university academics). The interactions and collaboration of the community were learning-centred with a common theme of improving participation and performance in mathematics at their respective schools.

In our deliberation with teachers in the LEDIMTALI project, we found that they valued the following features of a PLC: a common vision for student learning, collective solution-seeking, collaborative construction of teaching and learning resources, sharing private practice through collegial reflection and feedback, and shared and supportive leadership.

Wang (2015) suggests that researchers should study the differences that manifest between PLCs operating in high performing educational systems and PLCs operating in developing contexts. An example of one difference suggested is that collective inquiry as a feature of a PLC may manifest in high performing educational systems as collaborative knowledge construction, whereas in a developing educational system it may be noticeable as collaborative solution seeking.

Features of a PLC that account for professional learning

This chapter proposes two aspects of community that account for professional learning: The notion of 'a micro-climate of commonality' (McMillan & Chavis, 1986; Rovai, 2000); and the notion of relational agency (Edwards, 2007). We now consider the theoretical underpinnings of these two constructs.

From the broader study we have done, it became obvious that the concept of community does not adequately account for teacher perceptions about collaboration and mutual support in a PLC. Besides the fact that the community concept is very broad and sometimes vaguely defined, it can also be that issues such as pseudo-community, contrived collegiality and conflict (Achinstein, 2002) dilute the socio-cultural epistemological perspective as a basis for collaborative learning. Therefore, the notion of a micro-climate of commonality (McMillan & Chavis, 1986) is advanced as a mechanism to explain how community facilitates professional learning.

A micro-climate of commonality is defined by the following elements: membership, influence, the integration and fulfilment of personal needs, and a shared emotional connection. Table 8.1 explicates these elements.

TABLE 8.1   A micro-climate of commonality

Elements that make up a micro-climate of commonality	Explanation: teacher experiences or affordances related to community.
Membership	A feeling of belonging and personal relatedness to other colleagues
Influence	Opportunities to make a contribution or a difference in the lives of other members
Integration and the fulfilment of personal needs	Accessibility to support structures and resources
Shared emotional connection	A commitment to the community and a belief that members need each other and will be there for each other

Source Adapted from the work of McMillan and Chavis (1986) and Rovai (2000)

Riveros, Newton & Burgess (2012) express a concern that PLCs have become instrumentalist in orientation only focusing on how to transform teaching practice in order to enhance student learning. They lament that not much attention is being given to teacher learning and teacher knowledge in the PLC structure. This is the gap that we attempt to address by introducing the concept of relational agency in the context of mathematics teacher professional learning.

Relational agency

The concept of 'relational agency' emanates from the work of Edwards (2005, 2007) and is a useful construct to analyse and interpret teacher collaboration. Relational agency refers to the capacity of the participants to seek assistance as well as to offer assistance based on the relationship that was established in the PLC. Edwards describes relational agency as follows:

> It is a capacity which involves recognising that another person may be a resource and that work needs to be done to elicit, recognise and negotiate the use of that resource in order to align oneself in joint action on the object. (Edwards, 2005:173)

Relational agency, therefore, has a dynamic quality that emphasises reciprocity and mutual strengthening of competence and expertise to enhance the collective competence of a community.

Relational agency may also account for the teachers' enthusiastic participation in the project. This supposition is predicated on the belief that the efficacy of collaborative approaches towards the appropriation of new knowledge, skills and attitudes is supported by the social network theory (Daly, 2010; Moolenaar, 2012). This theory posits that social structure and the network of interpersonal relationships serve as an affordance for collaborating participants to access resources made available through the network (Moolenaar, 2012). In this regard, we see resources as clustered into four categories, namely financial, physical, human and information (Sugarman & Martin, 2011).

Relational agency may become apparent in a collaborative setting in several ways. Some of the ways it may be identified as operational in a PLC is when teachers:

Ø   recognise each other as a resource for professional learning;

Ø   share expertise and teaching resources;

Ø   are willing to act as a resource for others in their quest to pursue professional learning; and

Ø   display the propensity as well as the capacity to collaborate.

Relational agency can be enacted in two ways: firstly, through direct participation in PD activities during workshops, and secondly, through indirect participation in PD through collegial interactions at the same school or across schools. Relational agency enacted as indicated here connects the individual to the collective in a PLC. As such, participants take responsibility for each other's learning through reciprocity and strengthening of competence. Relational agency also has the potential to configure interactions beyond the scope of a PLC into powerful professional learning opportunities, such as informal discussions with a colleague over tea or conversations with learners during a classroom episode (Riveros et al., 2012).

The role of relational agency

Hargreaves (2007) argues that strong and sustainable PLCs are characterised by strong cultures of trusted colleagues who value each other personally and professionally. As such, this chapter suggests that one of the most important attributes of such a strong collegial culture is the notion of relational agency. Hence, within the community dimension of a PLC the role of relational agency is to (a) acknowledge one another's ideas and dignity, (b) believe in each other's ability and willingness to fulfil professional responsibilities, (c) care about each other both professionally and personally, and (d) trust one another and reciprocate support in pursuing professional learning (Peretti, 2009).

Figure 8.1 visually represents the interplay between the constructs of community, micro-climate of commonality, relational agency and PL.

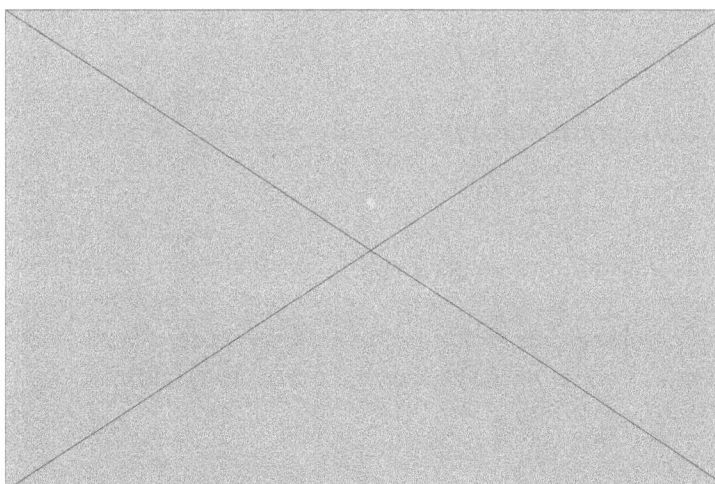

FIGURE 8.1 Connection between micro-climate of commonality, relational agency and professional learning

Figure 8.1 illustrates that professional learning is facilitated by the attributes afforded in this diagram. The attributes are explicated in the clouds on the right-hand side of the diagram. The facilitation of professional learning is occasioned by the realisation of relational interdependency, which is guided by reciprocity, lateral accountability and distributed leadership. This relational interdependency is rooted in the way commonality functions in the micro-climate of commonality. In this way, a micro-climate of commonality accounts for the socio-cultural epistemology through which teachers may experience growth in knowledge and skills.

This relational interdependency presupposes a community of learners in which PLC participants experience facets of caring, mutual support, acceptance, social relationships and engagement, friendship, respect and psychological safety (Orland-Barak, 2017). Hence, a micro-climate of commonality is attributed with a set of dynamic relationships embedded in the contextual and historical composition of the PLC.

Analysis and discussion

The research for this chapter is grounded in the qualitative research tradition and follows a phenomenological approach (Flick, 2018). Data gathering methods employed individual as well as focus group interviews.

The LEDIMTALI PLC consists of ten schools, and about 45 teachers participated in its regular activities. During the third year of the project, twenty-one teachers were purposively selected and invited to participate in the study. All of the participants in the study were actively involved in the PLC for at least three years. Of the teachers participating in the focus groups and individual interviews, nine were identified to participate in the individual interviews and the rest participated in the focus group interviews.

The sample included three male and six female teachers. Three of the teachers interviewed held senior positions, namely that of a head of department. Between them, the interviewees held 125 years of experience, or on average approximately 18 years of experience in the teaching of mathematics from Grades 9-12. The size of their subject departments averaged five teachers per school. Fictitious names are used for teachers in the section under discussion and findings.

The interviews were audio-recorded and transcribed verbatim for analysis of the data. A thematic data analysis approach was employed to analyse the interviews (Lacey & Luff, 2007).

The research for this chapter was driven by two questions:

1. What were the mathematics teachers' perceptions of working collaboratively in the context of a PLC?
2. What are the key affordances for mathematics teacher professional learning of such a collaborative approach?

These two questions will frame the subsequent discussion.

Mathematics teachers' perceptions of working in a collaborative context

Analysing teachers' responses pointed towards three crucial constructs that form the backbone of professional learning in a collaborative setting, namely community, a micro-climate of commonality, and relational agency. These findings are aligned with the theoretical framework illustrated in Figure 8.1 on the previous page. Some of the teacher responses are cited below:

Focus group 1   For us community implies that teachers from all walks of life coming together learn from each other by sharing their experiences and challenges. They learn together and learn from each other. They share resources. Working together with experts of the university is also a good thing because they bring new knowledge and research to us, but they are also teachers and so there is no difference between us, it is just that we teach at a different level.

Participants also alluded to the opportunity the PLC affords community members to seek and obtain assistance from other participants based on friendships that developed in the PLC (Biesta, Priestley & Robinson, 2017). Eleanor used the concept of a support group to describe relational agency and she articulated it in the following way:

Eleanor   We are looking for solutions in a collaborative way – we become like a support group where we can phone each other when we encounter problems in teaching.

Relational agency also obliges community members to assist when called upon to do so. In the words of one of the participants, this community consists of people who are willing to share their expertise:

Arlene   A support group means like you are there for each other and when they need you for assistance you do not say no.

The relational interdependency that teachers experienced is confirmed by the work of Rovai (2000:287) who posits: "[C]ommunity members develop a sense that they are not facing challenges alone and that they are needed and valued within the bounds of their community."

Key affordances for mathematics teacher professional learning

Professional development is part of the continual process of improving the practice of mathematics teachers who are employed as teachers in schools across the world. However, this idea has been overtaken by the notion of professional learning. It may seem like a mere change in terminology, but philosophically the difference is significant (Webster-Wright, 2009). Professional learning embodies many of the same ideas and goals of professional development. However, this notion of professional learning emphasises active as well as interactive learning by teachers. It signifies personal engagement rather than being subjected to adopting techniques and methodologies imposed by outside experts. For this learning process to become encultured in the teaching profession, the following key affordances play a significant role.

Collaborative solution seeking

Firstly, the data highlights that the LEDIMTALI PLC participants are committed to collaboratively seek solutions to the educational problems they encounter in their mathematics classrooms. For example, Eleanor felt that:

Eleanor   We are looking for solutions in a collaborative way.

Elvira also articulated this commitment in the following way:

Elvira   As teachers, we also saw this as an opportunity to improve our knowledge and teaching methods, since maybe our current teaching methods are outdated or do not work anymore.

A further example is Erik, who said:

Erik   We feel that it is important to continually improve ourselves in order to increase the learning of our children. Especially in a school like ours where the learners have a disadvantage and lags behind.

Secondly, for these teachers, collaborative solution seeking also means experimenting with new methods and giving feedback to each other as to how it works in their classes. For example, one interviewee reported as follows:

Interviewee   The spiral revision for me is one of the most outstanding features in LEDIMTALI. It works for me and helps to consolidate the learning of my learners.

In the LEDIMTALI PLC, the first classroom-based problem the group tackled was the failure on the part of students to do homework. Spiral revision was one of the solutions that emanated from this collective inquiry.

Deprivatised practice

Deprivatisation is a characteristic of professional learning communities as defined by Louis and Kruse (1995). This feature of PLCs relates to the practice of teachers visiting each other's classes, providing non-evaluative feedback to each other and developing a culture of trust and collaboration. In general, teachers are mostly engaged in image management and thus are reluctant to open their doors to their colleagues. In the research for the first author's PhD thesis, one of the interviewees commented:

Safety, both psychologically and physically is important for our wellbeing and confidence to allow a colleague into your class. Even just showing the video clips of some teachers teaching, can be threatening but if we trust each other we are not afraid to be observed and reflect on our practice.

In the LEDIMTALI PLC, there was a concerted effort to encourage peer visitation for the purpose of giving non-evaluative feedback to each other. However, from the interview data, it was apparent that organisational arrangements at the school level remain a barrier because of teachers' full teaching schedules and after-hours commitments. This was explained by Arlene, one of the teachers interviewed:

Arlene   No, it was not possible to visit other colleagues' classes. We were trying to do that, but the problem is the timetable; time does not allow us to do that. We have many classes ourselves which we have to teach.

The one way in which deprivatisation of practice was facilitated in the LEDIMTALI PLC was to video record lessons (with the permission of the teacher) and then review these lessons in a safe environment. Firstly, this video recording was viewed by a facilitator and the teacher alone and the two reflected on the content of the recording. If the teacher concerned felt comfortable, the video was shared in a larger group setting so that more PLC participants could experience the way of doing peer observation and engaging in a non-evaluative reflection of classroom practice. In this way, more teachers were encouraged to participate in this PLC.

Dealing with diversity

Diversity denotes the variety of differences between individuals in an organisation. This includes race, gender, ethnic groups, age, religion, sexual orientation and citizenship status. Diversity not only includes how individuals identify themselves but also how others perceive them. In a project such as LEDIMTALI, some of these issues are easily dealt with given the history of South Africa and the dispensation introduced after 1994. However, within any learning community, the diverse views of participants need to be managed.

Concerning dealing with diverse views or opinions amongst participants, Andrew, one of the participants voiced the general feeling amongst the group:

Andrew   LEDIMTALI is welcoming and accommodative When there is a divergence of opinion or disagreement the leader let us discuss in order to reach a common understanding.

In this, he is supported by the views of many other participants. This is what was said during the focus group interviews:

Focus group interview   A PLC is where teachers from different schools get together to share ideas. We work as a team In the PLC we are treated as equals and this leads to openness and trust. Even if we differ, we see that as an opportunity to learn.

This last comment indicates that the culture that developed in the PLC was robust enough to mediate diversity issues. Another indicator of accepting diversity in terms of language, race and culture is captured in Erick's statement:

Erick   I have never had problems with diversity because South Africa is a diverse country. It is part of everyday living. This actually contributes to enriching ourselves.

Conclusion

The aim of this chapter was to draw upon socio-cultural perspectives to explore the role of relational agency and how it impacted on mathematics teachers' professional learning in the context of a PLC. This chapter attempted to facilitate a better understanding of how professional learning may be facilitated through the affordances within a collaborative context and the role of relational agency in such a context to enhance professional learning. In relation to the notion of "a micro-climate of commonality" as espoused by Rovai (2000), the other attributes noted by McMillan and Chavis (1986) become relevant to the experiences of the participating teachers. Through a shared process involving increased trust and togetherness, the teacher, as well as the teacher educators, became empowered in terms of their collective identity and agency. (Hökkä, Vähäsantanen & Mahlakaarto, 2017).

Furthermore, this chapter found that mathematics teachers' perception of community is coloured by a sense of interdependence and mutual obligation. Adapting the suggestions forwarded by Palloff and Pratt (1999), the following could be the outcome of an intentional intervention to facilitate the growth and development of a micro-climate of commonality and relational agency amongst teachers:

Ø   Socially constructed meaning evidenced by agreement or questioning, with the intent to achieve agreement on issues of meaning.

Ø   Sharing of resources among mathematics teachers.

Ø   Expressions of support and encouragement exchanged between teachers.

Ø   A willingness to critically evaluate the ideas of others.

Finally, this chapter elucidated the efficacy of relational agency as an analytic to understand the inner workings of professional learning communities with the emphasis on community, in the context of CPD of mathematics teachers.

References

Abu-Tineh, A.M. & Sadiq, H.M. 2017. Characteristics and models of effective professional development: the case of schoolteachers in Qatar. *Journal of Professional Development in Education*, 43(1):1-12.

Achinstein, B. 2002. Conflict amid community: The micro-politics of teacher collaboration. *The Teachers College Record*, 104(3):421-455. https://doi.org/10.1111/1467-9620.00168

Biesta, G.; Priestley, M. & Robinson, S. 2017. Talking about education: exploring the significance of teachers' talk for teacher agency. *Journal of Curriculum Studies*, 49(1):38-54. https://doi.org/10.1080/00220272.2016.1205143

Daly, A.J. (ed.). 2010. *Social network theory and educational change* (Vol. 8). Cambridge, MA: Harvard Education Press.

Dufour, R. 2004. What is a 'professional learning community'? *Educational Leadership*, 61(8):6-11.

Edwards, A. 2005. Let's get beyond community and practice. The many meanings of learning by participating. *Curriculum Journal*, 16(1):49-65.

Edwards, A. 2007. Relational agency in professional practice: A CHAT analysis. *Action: An International Journal of Human Activity Theory*, 1:1-17. https://doi.org/10.1080/0958517042000336809

Eliahoo, R. 2017. Teacher educators: Proposing new professional development models within English further education context. *Journal of Professional Development in Education*, 34(2):179-193. https://doi.org/10.1080/19415257.2016.1178164

Grossman, P.L.; Wineburg, S. & Woolworth, S. 2001. Toward a theory of teacher community. *Teachers College Record*, 103(6):942-1012. https://doi.org/10.1111/0161-4681.00140

Hargreaves, A. 2007. Sustainable professional learning communities. In: L. Stoll & K.S. Louis (eds.), *Professional learning communities: Divergence, depth and dilemmas*. Berkshire, England: Open University Press. pp.181-195.

Hökkä, P.; Vähäsantanen, K. & Mahlakaarto, S. 2017. Teacher educators' collective professional agency and identity – Transforming marginality to strength. *Teaching and Teacher Education*, 63:36-46. https://doi.org/10.1016/j.tate.2016.12.001

Kennedy, E. 2006. Models of continuing professional development: A framework for analysis. *Journal of In-Service Education*, 31(2):231-250. https://doi.org/10.1080/13674580500200277

Lacey, A. & Luff, D. 2007. *Qualitative research analysis*. Nottingham and Sheffield: NIHR RDS for the East Midlands/Yorkshire & the Humber.

Lave, J. & Wenger, E. 1991. *Situated learning, legitimate peripheral participation*. Cambridge, UK: University of Cambridge Press. https://doi.org/10.1017/CBO9780511815355

Louis, K.S. & Kruse, S.D. 1995. *Professionalism and community: Perspectives on reforming urban schools*. Thousand Oaks, CA: Corwin.

McMillan, D.W. & Chavis, D.M. 1986. Sense of community: A definition and theory. *Journal of Community Psychology*, 14(1):6-23. https://doi.org/10.1002/1520-6629(198601)14:1<6::AID-JCOP2290140103>3.0.CO;2-I

McNichol, J. 2013. Relational agency and teacher development: a CHAT analysis of a collaborative professional inquiry project with biology teachers. *European Journal of Teacher Education*, 36(2):218-232. https://doi.org/10.1080/02619768.2012.686992

Moolenaar, N.M. 2012. A social network perspective on teacher collaboration in schools: Theory, methodology, and applications. *American Journal of Education*, 119(1):7-39. https://doi.org/10.1086/667715

Orland-Barak, L. 2017. Learning teacher agency in teacher education. In: D. Clandidin & J. Husa, *The SAGE handbook of research on teacher education*. Helsinki: University of Helsinki.

Palloff, R.M. & Pratt, K. 1999. *Building learning communities in cyberspace* (Vol. 12). San Francisco: Jossey-Bass.

Peretti, M.Q. 2009. *Teachers' perceptions of professional learning communities: Measuring the impact of administrative participation.* PhD thesis, Walden University, ProQuest Thesis Publishing.

Riveros, A.; Newton, P. & Burgess, D. 2012. A situated account of teacher agency and learning: Critical reflections on professional learning communities. *Canadian Journal of Education/ Revue comedienne de l'éducation*, 35(1):202-216.

Rovai, A.P. 2000. Building and sustaining community in asynchronous learning networks. *The Internet and higher education*, 3(4):285-297. https://doi.org/10.1016/S1096-7516(01)00037-9

Sugarman, J. & Martin, J. 2011. Theorizing relational agency. *Journal of Constructivist Psychology*, 24(4):283-289. https://doi.org/10.1080/10720537.2011.593455

Wang, T. 2015. Contrived collegiality versus genuine collegiality: Demystifying professional learning communities in Chinese schools. *Compare: A Journal of Comparative and International Education*, 45(6):908-930. https://doi.org/10.1080/03057925.2014.952953

Webster-Wright, A. 2009. Reframing professional development through understanding authentic professional learning. *Review of educational research*, 79(2):702-739. https://doi.org/10.3102/0034654308330970

Westheimer, J. 1999. Communities and consequences: An inquiry into ideology and practice in teachers' professional work. *Educational Administration Quarterly*, 35(1):71-105. https://doi.org/10.1177/00131619921968473

| O 9 |

AN ETHNOMETHODOLOGICAL ANALYSIS OF CANDIDATES' WORK IN HIGH-STAKES MATHEMATICS EXAMINATIONS AND ITS POSSIBLE USE IN CPD INITIATIVES

Marius Simon, Cyril Julie & Lorna Holtman

Introduction

The studies on learners' work on errors and misconceptions are primarily done from a cognitivist perspective and what might be termed a 'mathematical correctness comparative' perspective. The former is elaborated by Davis (1984), whose project was to develop a language of description to deal with both incorrect and correct work produced based on cognition. Marilyn Matz viewed the approach as explaining the "uniformity of the answers produced – both correct and incorrect" in mathematical problem-solving. She exemplifies the approach with many examples of errors learners make in algebra. For, example, she demonstrates how through generalising the distributive law, A(B + C) = AB + BC, learners would extrapolate this to the simplification of $\sqrt{A^2 + B^2} = A + B$ (Matz, 1980:112).

The 'mathematical correctness comparative' perspective is simply the way of assessing learners' responses to problems by comparing their work to the mathematically correct procedures to arrive at the answers. For high-stakes examinations, these mathematically correct methods are normally contained in a memorandum or scheme of marking. The scheme of marking contains as many possible correct solution methods and where learners produce further correct solution paths these are added to the memorandum of marking. For example, the memorandum of marking for NSC Mathematics examination for 2014 had three different ways of arriving at the required response for the item "Solve for x: $2^{x+2} + 2^x = 20$." (Department of Basic Education, DBE, 2014a:3). Figure 9.1 gives the scheme of marking.

1.1.3	$2^{x+2} + 2^x = 20$	✓ common factor/*gemeen. faktor*	
	$2^x(2^2 + 1) = 20$	✓ simplification/*vereenvoudig*	
	$2^x = \frac{20}{5}$		
	$2^x = 2^2$	✓ answer/antwoord	(3)
	$\therefore x = 2$		
OR/OF	$2^x.2^2 + 2^x = 2^2.5$	✓ common factor/*gemeen. faktor*	
	$2^x(2^2 + 1) = 2^2.5$	✓ simplification/*vereenvoudig*	
	$2^x.5 = 2^2.5$		
	$\therefore x = 2$	✓ answer/antwoord	(3)
OR/OF	$4.2^2 + 2^x = 20$		
	$5.2^2 = 20$	✓ $5.2^2 = 20$	
	$2^x = 4 = 2$	✓ $2^x = 4$	
	$\therefore x = 2$	✓ answer/antwoord	(3)

FIGURE 9.1 Memorandum for x: $\mathbf{2^{x+2} + 2x = 20}$

Source *DBE (2014b:30)*

Instructions accompanying a memorandum, such as the one in Figure 9.1, are that positive marking must be applied. This means that where a candidate's solution path breaks down, only that part of the work is penalised. If the breakdown is of such a nature that the intent of the item is not severely compromised then the rest of the work is marked. In the worst case, their procedures are also sometimes evaluated as incorrect, as was the case for the item discussed here. The diagnostic report on the examination states:

> When solving exponential equations, candidates displayed no understanding of the basic exponential laws and struggled to manipulate expressions involving the

addition and subtraction of exponents. Some candidates used incorrect mathematics and arrived at the correct answer.

$2^{x+2} + 2^x = 2^4 + 2^2$
$x + 2 + x = 4 + 2$
$\quad x = 2$

They were awarded no marks for this response. (DBE, 2014c:112)

What is lost sight of in the above comment is that there exists a special set of exponential equations with the same structure as the one under scrutiny which can be solved the way some candidates did it.

The perspective taken in this study is a quasi-ethnomethodological one where the focus is not on the correctness or incorrectness of the responses but rather on teasing out how the high-stakes examination context, with the candidate as an inextricable part of this context, contribute towards the production of the responses by the candidates.

The next section presents a brief description of the theoretical constructs of the chapter. This is followed by an analysis of actual learners' responses to items on trigonometry in a high-stakes examination from the perspective of ethnomethodology. The final section reflects on the use of such analyses in CPD for high-school mathematics teachers with the aim of improving achievement in high-stakes mathematics examinations.

Theoretical machinery

As mentioned previously, an ethnomethodological perspective underpins the analysis. Ethnomethodology is the study of how ordinary people use common sense knowledge, procedures and considerations to gain an understanding of everyday situations (Garfinkel, 1967). Ethnomethodology's objective is to extract the social facts from practical social actions. The process of extracting social facts from practical social actions is clarified in Durkheim's (1938) description of sociological studies when he established the objective reality of social facts as sociology's fundamental principle. Garfinkel's (1967) re-specification of this gave ethnomethodology its particular foundation in the field of sociological research.

> Ethnomethodology sees the objective reality of social facts, as how it is every society's locally, endogenously produced, naturally organised, reflexively accountable, ongoing practical achievement, being everywhere, always, only, exactly and entirely, members' work, with no timeouts, and with no possibility of evasion, hiding out, passing, postponement, or buy-outs, is thereby sociology's fundamental phenomena. Every topic of detail and every topic of order is to be

> discovered and is discoverable and is to be re-specified and is respecifiable as only locally and reflexively produced naturally accountable phenomena of order. (Garfinkel, 1991:11).

What Garfinkel (1967) proposes here is that all the properties of social order are made visible locally with the result that social scientists and ethnomethodologists can see what participants do in social settings. The organisation of social order occurs in its natural setting (Liberman, 2012). This perspective comprises ethnomethodology and constructs from the social study of science. The constructs of importance are lived and reported work, reflexivity, the mathematically historicised-self, ethnomethodological indifference, human and disciplinary agencies with a particular mention of resistance and accommodation.

Lived and reported work

A defining feature of ethnomethodology and constructs from the social study of science is that the researching of the production of any artefact is done in situ. Livingston (1986) calls this work the 'lived work'. He distinguishes the 'lived work' from the 'account' of the work, which is the clean, orderly presentation of the 'lived work', such as articles in journals or books. Livingston followed this line of analysis to expose how mathematical proofs are constructed in situ using the theorem: The angle at the centre of a circle is twice the angle subtended by the same chord at the circumference (Livingston, 1987:116). Roth (2013) follows a similar procedure to reveal the emergence of the proof that the sum of the angles of a triangle equals 180°. In both these instances, the teasing out of the in situ work is not done by observing the actual work being done in the context of doing the work. Rather, the setting is reconstructed in terms of how the work proceeds in a navigational space, such as a chalkboard or on paper.

A slightly different approach followed is the direct observation of the work. This approach normally includes a video recording of the happenings at the worksite and, where possible, accompanying documents that can trace the progress of the 'lived work'. Julie (2003) followed this approach when he analysed the work of practicing teachers constructing a mathematical model. Greiffenhagen (2008) also used this approach when analysing how mathematics unfolds in lecture situations.

It is, however, not always possible to have access to the 'lived work' through direct observation or the reconstruction of the navigational space based on the experience of the work by the analyst. In such cases, communicative documents, such as e-mails between the producers of the work, are used to trace the lived work. Merz and Knorr-Cetina (1997) used such documents to study how theoretical physicists at the CERN laboratory accomplished results to problems they encountered in their work. Andrew Pickering (1995) used the actual detailed reflective account by Hamilton to analyse the path followed by Hamilton in his construction of quaternions.

Documents accompanying the video-recordings (if done) and normally the so-called 'rough notes and scribblings' of the producers of the work are some of the primary sources used to tease out the 'lived work'. It is widely known that 'rough notes and scribblings' of mathematical work normally end up in wastepaper baskets and are not readily available. If these are available, they are used in conjunction with the 'work account' to tease out the navigational route followed to render the 'work account'.

Reflexivity

In terms of mathematical work, the meaning of any work account or action including the 'lived work' becomes evident by investigating the various ways that participants solve mathematical problems. Hence, the visible accomplished work is reflexive and is shaped by localised settings. Reflexivity, in this instance, refers to the "self-explicating property of ordinary actions" (ten Have, 2004:20). The reflexive nature of all interaction refers to the ability of practices to both describe and constitute a social framework (Coulon, 1995). In this sense, reflexivity is not a conscious process. Participants (candidates) generate a constantly emerging social reality through the dialectic production of meaning in interaction. In the process of seeking a solution to a mathematical problem, the objective, which is set by the question paper as problem text, drives the action taken from it. The practical social action exists in all thinking processes, and hence also in the choice of mathematical skills selected to accomplish what was set out for the candidate to achieve. Garfinkel (1991:4) sees these repeated scenes of human social action as 'reflexively accountable'. In a simplified manner, this 'reflexively accountable' with regard to dealing with mathematical problems in the high-stakes examination setting can be viewed as "the next steps to be executed in the solution-seeking pursuit and are guided by 'looking back' to prior executed steps."

The mathematically historicised self

The mathematically historicised-self includes all school mathematics that the candidate was exposed to in the school and non-school contexts in all forms. The candidates take into account that which is known in terms of ways of doing mathematics in certain ways – a set of skills that were developed over the years of schooling in the subject. During schooling, the candidates are exposed to multiple forms of tests and examinations to consider their progress and whether to secure or halt their promotion. These school-based examinations are normally modelled on the NSC examination. It is thus understandable that the candidates enter the examination room not only with the knowledge they will be tested on but with all school mathematical knowledge and experiences known to them. This chapter shows how the mathematically historicised-self comes into play in the form of little reminders candidates use to build up their understanding of the mathematical problems put to them to solve.

Ethnomethodological indifference

Ethnomethodological indifference is an instructive way of working as it involves deliberately avoiding being judgmental with respect to whether ongoing productions are correct or incorrect. This is a fundamental principle of ethnomethodology in that decisions of constituents of phenomena are neither made in advance nor are they based upon prior formal analytical studies (Garfinkel 2002:171). Relevant to this chapter, the central idea of the use of ethnomethodological studies is to provide an alternative procedural description of the phenomena of order and method regardless of whether the produced responses are correct or incorrect. The analysis of the responses of the candidates is to, in a post-hoc manner, indicate how they went about doing the work and interpreting the order of this doing. Thus the pursuit is about how claims are accepted or discredited and whether the shared assumptions or tacit knowledge underlying of what is accomplished – a defining feature of ethno-methodological analysis.

Human and disciplinary agencies

Pickering (1995) captures a tangled relationship between human and material agencies. This relationship is counter to Collins and Yearley's (1992) defence of the humanist positioning of the traditional sociology of scientific knowledge, which gives preference to the human subject through an unbalanced distribution of agency to humans and none to the material world. Like other practice theorists, Pickering defines entanglement between human and material agencies as concepts of practice which involves cultural and historical activity. He sees this relationship as "the work of cultural extension and transformation in time" (Pickering, 1995:5).

According to Pickering, disciplinary agency is the counterpart of machine agency when non-machinic work is at stake. However, he proposes that the ongoing social routines of human agency are located in disciplinary agency consisting of resistances within conceptual practices (Pickering, 1995:29). Human agency accompanies discursive structures in discursive practices as does material agency in material practices. Disciplinary agency is thus the counterpart to material agency when it comes to the practice of solution-seeking of mathematical problems (Pickering, 1995). It consists of historically established, routinised and structured operational techniques that are applied and used in a mathematical domain. Performing such techniques does not interfere with the goals and intentions of the human practitioners. During these times, the creativity of the scientist is passive and the disciplinary agency in an active mode.

In mathematics, one can argue that the tools or non-human agencies and the role these play in the production of mathematics are not easily recognisable as the disciplinary agency of any mathematical work. However, it can be accepted that the final answer is when humans give up control to the routine ways of reacting to some mathematical problem. This chapter recognises the composite relationship of human and disciplinary agency and draws further insight from what Holland, Lachicotte, Skinner and Cain (2001:279) state regarding agency: that it is not just individual, but rather that "agency lies in the improvisation that people create in response to particular situations". These responses people create can only be observed and reacted upon. The product of this 'mangle of practice', as Pickering (1995) calls it, constitutes the answer texts produced by candidates in the high-stakes Grade 12 NSC Mathematics examination.

The relevance of agency with regards to this chapter reveals the capacity of candidates to do things, or in this case to produce mathematics. Disciplinary agency has the capacity to bring forth the unexpected, which in turn exerts agency that the candidate can observe and react to (e.g., checking an answer; abandonment or adaptation). This notion of acting and reacting is what Pickering (1995) refers to as the dialectic of resistance and accommodation.

Resistance and accommodation

Pickering defines the "occurrence of a block on the path to some goal" as resistance (Pickering, 1993:569). This resistance is accommodated by different interactions. Pickering (1993) sees accommodation as some tentative human approach to circumvent the obstacles. This dialectic mechanism captures how sense-making emerges over time in a social setting.

If mathematics production is to be fully understood, the tools used in mathematical activity are not to be reduced to avoidable phenomena. In the instance of the work reported in this chapter, the situated environment has many elements that are part of the existing social order. It is important to note that these tools do not operate independently, but instead, require a skilled person to channel their agency in the desired direction. This implies that the skilled operator, in this case, the candidate, and the non-human agency come together as a single unit in the production of mathematics in the situated environment. Furthermore, in elaborating on the co-existence of human and disciplinary agencies by referring to them as scientific practices, Pickering (1995) views science as practice and culture. Thus, what exists in this social order is the sociological engagement of human agency with disciplinary agency coupled to the scientific dialectic of resistance and accommodation. Through the dialectic resistance and accommodation and the ethnomethodological constructs, the textures of the candidates' ways of working in their pursuit of an objective or solution are made visible.

Analysis procedure and data

In this study, the analytical focus relies on Garfinkel's (1967:78) documentary method. According to Dufour (2013), a documentary analysis seeks to examine the tacit knowledge underlying the textures of participants' ways of working when doing mathematics. Pickering (1995:22) studied historical documents on Hamilton's construction of quaternions to account how human and disciplinary agency drive the meaning-making process through the "the dialectics of resistance and accommodation".

The use of ethnomethodological principles and the dialectic of resistance and accommodation influenced the current analysis. Garfinkel (1967:8) refers to the "rational character of [the] actual". The 'actual' for this chapter is embedded in the text that learners produced and the 'rational character' of it is inferred from the entire text – scribblings, strikethroughs, and such like – that were available for scrutiny.

Central to the approach adopted in this chapter is the local natural setting in which candidates interacted with the problems in the high-stakes examination. This study is premised on the notion that any local natural setting influences the production of mathematical work, whether in the mathematics classroom or an examination. Julie (2015) describes the setting of the high-stakes NSC examination as one consisting of the examination text comprising the problems to be solved; the resources, such as the sheet with formulae accompanying the examination text; a non-programmable and non-graphing calculator; the answer book on which both the rough work and final responses to the problems are recorded; and timing devices used by the candidates and the invigilators to remind candidates about their time already taken for completing the time-restricted examination. This context structure drives the resolution-seeking pursuit for responding to the problems posed in the examination texts.

The data consisted only of the answer texts or scripts from the 2012 NSC Mathematics examination. The work is from candidates from schools that participated in the LEDIMTALI project. The sample is thus an opportunistic one. Mugo (2002) explains an opportunistic sample as a form of purposeful sampling where information-rich cases for in-depth study are selected. The opportunistic sample selection allows the researcher to take advantage of unexpected flexibility.

Results

The presented results are guided by Livingston's call for "a vocabulary that provides descriptive insight into the lived, perceived details and reasoning of mathematical discovery work" (Livingston, 2006:63). This points in the direction to the paucity of settled constructs from ethnomethodology to describe the work of the production of mathematics. What do exist are different locally specific categories constructed by researchers. The results presented and discussed in this section are anchored around the notion of abandonment. Abandonment is when a particular resolution-seeking pursuit route is started and then partly or wholly discarded due to some triggering after some realisation that the solution route is not going in the required direction (Figure 9.3, 9.4, 9.5, 9.7, 9.8, 9.10 and 9.11 are reproduced from the actual scripts of the candidates).

Abandonment after a nearly complete attempt

Abandonment after a nearly complete attempt occurs when the candidate follows a solution path almost up to a final answer and then relinquishes the chosen path.

Figure 9.2 presents the problem of interest being focused on. The category under discussion focuses on 8.1.4, but other parts of the question play a part in the production of the response to 8.1.4.

QUESTION 8

Answer the question WITHOUT using a calculator.

8.1   The point P(k ; 8) lies in the first quadrant
        such that OP = 17 units and TÔP = a
        as shown in the diagram alongside.

8.1.1   Determine the value of k.   (2)

8.1.2   Write down the calue of cos a.   (1)

8.1.3   If it is further given that $a + ß = 180°$, determine cos ß.   (2)

8.1.4   Hence, determine the value of $\sin(ß - a)$.   (4)

FIGURE 9.2   November 2012 NSC Mathematics Paper 2, Question 8

$\sin(ß - a)$		Line 1
$\sin(ß - a)$	$= \sinß - a.\cos a \cdot \cosß.\sin a$	Line 2
	$= \sin \frac{-17}{8}$	Line 3
$\sin(ß - a) = \sin ß.\cos a - \cosß.\sin a$		Line 4
	$= \sin \frac{19}{8} \cdot \frac{8}{17} - \frac{-17}{8} \cdot \frac{8}{17}$	Line 5
	$= \frac{19}{17} - 1$	Line 6

FIGURE 9.3   Abandonment after a nearly complete attempt

Figure 9.3 is a response to Question 8.1.4, Line 1 indicates that the candidate first wrote down the question. This served as a kind of reminder of "what must be determined". The expanded compound angle formula showed a transcription error. The information sheet gave the compound angle formula in the form of sin $(a - ß) = sin\ a.cosß - cosa.sinß$ whereas for the question $sin(ß - a)$ was used.

In line 2 of Figure 9.3 the 'what must be determined' is written down again in the expansion of $sin\ (ß - a) = sin\ ß - a.cosa - cosß.sina$ The candidate wrote $sin\ ß - a$ instead of only writing $sin\ ß$.

Upon writing $sin\ -\frac{17}{8}$ the pursuit is abandoned, which is indicated by the work being crossed out. This is evidence that there was resistance and some form of accommodation by continuing with a somewhat different solution path.

A new attempt is started in line 4 with the correct expansion for the compound angle formula. Values for trigonometric ratios are substituted but no disciplinary agency is exerted by, for example, $-1 \leq sin\ x \leq 1$.

Figures 9.4 and 9.5, which are responses to Questions 8.1.1 and 8.1.2, show the candidate's calculations for the values used in Figure 9.3.

8.1

8.1.1

r^2	$=$	$x^2 + y^2$	Line 1
$(17)^2$	$=$	$x^2 + (8)^2$	Line 2
15	$=$	x	Line 3
$\therefore$ k	$=$	15	Line 4

FIGURE 9.4   Question 8.1.1. Candidate determining the value of 'k'

8.1.2

$cos\ a$	$=$	$\frac{y}{r}$	Line 1
$cos\ a$	$=$	$\frac{8}{17}$	Line 2

FIGURE 9.5   Question 8.1.2. Candidate capturing the value for $Cosa$

Ultimately the expression $\frac{19}{17} - 1$ is arrived at (line 6 of Figure 9.3) and the entire pursuit is abandoned indicated by the non-simplification of the numerical expression.

Abandonment of a solution

The abandonment of a solution category has two sub-categories and they are discussed next.

Abandonment and the construction of a new problem by 'proving the given'

This way of working refers to the completion of a pursuit for a solution to a mathematical problem and then the abandonment of all the work, the construction of a different new problem and the pursuit of the solution of this newly constructed problem. Figure 9.6 presents questions, 11.1 and 11.2. The discussion focuses on 11.2.

QUESTION 11

ABCD is a parallelogram with AB = 3 units, BC = 2 units and ABC = θ for 0° < θ < 90°.

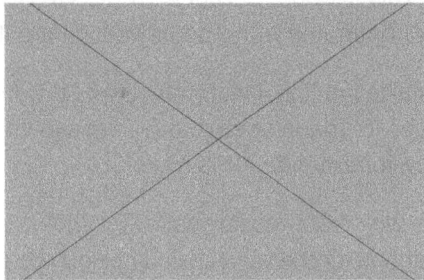

11.1　　　Prove that the area of parrallelogram ABCD is 6 sin θ.

11.2　　　Calculate the calue of θ for which the area of the parallelogram is 3√3 square units.

FIGURE 9.6　　　November 2012 NSC Mathematics Paper 2, Question 11

11.2　　　θ　　　=　　　$\frac{3\sqrt{3}}{3\sqrt{3}}$　　　Line 1　.

θ　　　=　　　60　　　Line 4

FIGURE 9.7　　　Abandonment of the first solution

In line 1 of Figure 9.7, the candidate wrote down the equation θ = 3√3 in an attempt to calculate the size of the angle. We infer that this was merely writing down the 'unknown' and the 'given'. This is common strategy learners are taught to solve problems prefaced by the word 'prove'. These are 'little reminders' written from the problem text to organise the solution-seeking pursuit. This way of working can be linked to the mathematically historicised-self particularly with respect to instructions given to learners when solving problems such as "write down the unknown and what is given".

Line 2 shows the introduction of *6sinθ* and its linkage to the given area of the parallelogram. This is to simplify the pursuit to get **θ** as the subject of the equation $\sin\boldsymbol{\theta} = \frac{\sqrt{3}}{2}$ (line 3). This is realised in line 4 through the presentation of $\boldsymbol{\theta} = \sin^{-1}(\frac{\sqrt{3}}{2})$. The solution format ($\sin^{-1}(\frac{\sqrt{3}}{2})$) demonstrates the agency of the historicised self. The way of instruction to find angles in trigonometry is to "press the button sin⁻¹". Line 6 renders **θ** = 60° as the final answer indicated by the drawing an arrow underneath it. The candidate abandons the entire pursuit by drawing two lines through it and writes the word 'CANCELLED' in between the two lines.

It now appears that the candidate realises that something is incorrect and there is deviation to solve a 'new' problem as illustrated in Figure 9.8.

11.2	Area of ΔABC	=	AB.BC sinθ	Line 1
		=	3 x 2 sin60⁰	Line 2
		=	$3\sqrt{3}$ units	Line 3

FIGURE 9.8   Newly constructed problem

The new pursuit shows the candidate's proof of what is given in the problem text in Figure 9.6. The newly constructed problem shows a pattern of work where, in the case of problems linked to 'areas' in trigonometry, some formula has to be found or proved. Thus the agency of the historicised self is exerted and there is no reversion to the problem text to ascertain what had to be done.

Abandonment and reconstruction of the same problem by 'proving the procedural objective'

This way of working is the abandonment of an obtained solution and following a different solution-seeking path through the reconstruction of the same problem with some adaptation.

Figure 9.9 gives the problem of interest.

9.2   Simplify without the use of a calculator: $\dfrac{\sin104°(2\cos^{2}15°-1)}{\tan38°.\sin^{2}412°}$

FIGURE 9.9   November 2012 NSC Mathematics Paper 2, Question 9.2

9.2

$$\frac{\sin104\,(2\cos^2 15° - 1)}{\tan38°.\,\sin^2 412}$$ Line 1

$$\frac{\sin104.2.\,(1 - \sin^2 15^0 - 1)}{\tan38°.\,\sin^2 52}$$ Line 2

$$\frac{\sin104.2.\,1 - \sin^2 15^0 - 1}{\dfrac{\sin38}{\cos38} - \cos^2 58}$$ Line 3

$$\frac{\sin104.2.\,1 + 3\sin^2 15^0}{\sin58.\,\cos58}$$ Line 4

$$\frac{\sin104.3 - 2\sin45}{\sin58.\cos58}$$ Line 5

1,24 Line 6

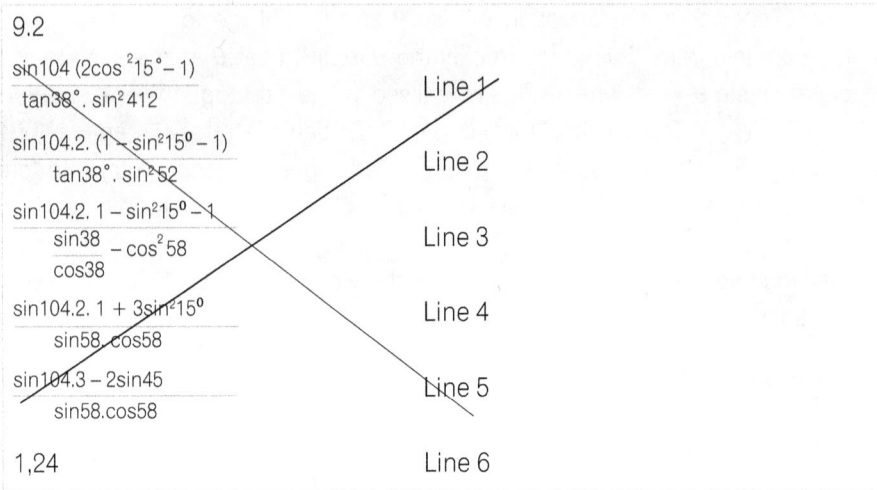

FIGURE 9.10   Abandonment of first solution-seeking path

Figure 9.10 is an example of the pursuit under discussion and presents the texture, which is the product of the way of working. The expression had to be simplified "without the use of a calculator". Line 2 of Figure 9.10 shows no change to the expression $sin104°$ and $tan38°$. However, $(2cos^2 15° - 1)$ is changed by substituting $cos^2 15°$ with $1 - sin^2 15°$, making use of the trigonometric identity relation of $cos^2 x + sin^2 x = 1$ to render 2.($1 - sin^2 15° - 1$). This exerted agency may be linked to the candidate's use of a structural sense which enables him or her to make better use of previously learned techniques. The presence of the brackets helped the candidate to 'see' structure. A feature of using structural sense is 'looking' before 'doing' – something that teachers commonly emphasise when reviewing previously learned work. The non-recognition of structural change in terms of the missing bracket might be caused by the continuation of work without 'noticing' the change that occurred.

The continuing navigation to work towards a solution sees the candidate substitute $sin^2 412°$ with $sin^2 52°$ using ($412° - 360°$). Line 3 of Figure 9.10 sees the removal of the brackets of the expression $2(1 - sin^2 15° - 1)$ to render $2.1 - sin^2 15 - 1$. Observably, there is no exertion of disciplinary agency for $2(1 - sin^2 15° - 1) = 2.1 - sin^2 15° - 1$ and resistance is not identified. The work continued to result in a calculator-generated final solution in line 6. The cancellation of the entire solution path was ostensibly triggered at this point by the problem texts "without the use of a calculator".

This is similar to Pickering's (1995) reasoning of Hamilton's construction of quaternions, finding resistance between an algebraic and geometric solution upon reaching the end of his pursuit. When encountering resistance, meaning something is not working, you start thinking of "making a new extension" (Marick, 2004), similar to Pickering's (1995) notion of accommodation to resolve a resistance.

The solution-seeking also reveals a form of reflexivity, combining the "where it will take me now" and as put by Coulon (1995:23), the "rationality of what we doing", in building up meaning "without the use of a calculator". The intention showed in the way of working is to produce a result "without the use of a calculator". It is almost as if the mathematical work in the abandoned pursuit and the objective set by the problem text with its underlying conceptual condition pre-determined the accounts of the new extension. This is shown in the handling of $sin104°$ to render $sin38°$. Figure 9.11 shows the adapted solution-seeking path.

9.2

$$\frac{sin104 \, (2(\cos^2 15° - 1)}{tan38°. \, sin^2 412} \qquad \text{Line 1}$$

$$\frac{sin76.2. + 2sin^2225}{tan^{sin38} . \, \frac{\cos^2 38}{\cos38}} \qquad \text{Line 2}$$

$$\frac{sin2.38.4sin^245}{sin38. \, \cos38} \qquad \text{Line 3}$$

$$\frac{sin38 \, \cos38.4sin^245}{sin38.\cos38} \qquad \text{Line 4}$$

$$4sin^245 \qquad \text{Line 5}$$

FIGURE 9.11   New extension

The new extension shows the candidate continuing with the same line of thinking, repeating some of the work produced in the abandoned pursuit. This shows that although the previous attempt was abandoned, some of the calculations were used without showing the steps; instead, they were copied into the new extension. The working towards cancellation is fuelled by the procedural instruction in the problem text, "without the use of a calculator". Line 2 of Figure 9.11, sees the simplification of $sin104°$ by using the supplementary angle relationship $sin(180° - 104°) = sin76°$ and in line 3, $sin76°$ is written as a double angle, $sin(2 \times 38°)$. The way of working shows the candidate produced the expression $\frac{sin38°\cos38°.4sin^245°}{sin38\cos38}$ rendered in line 4 of Figure 9.11. It is reasonable to infer that the sub-objective, triggered by the "without the use of a calculator" requirement, changed to a strategy where cancellation can be performed and special angles (45°) can be used. The final result is given as $4sin^2 45°$.

Concluding discussion

The results made visible candidates' ways of working in pursuance of solutions for trigonometry problems in a high-stakes examination from an ethnomethodological analysis perspective. It shows the exploration of these textures, exposing how candidates navigate their solution-seeking and discovery in pursuit of what was set out for them to accomplish in the examination. As stated above, it differs from the normal analysis of learners' work in high-stakes examinations. There is much value in this, and in fact, the success or not of learners in high-stakes mathematics examination depends on the way they deal with the problems presented in the examination.

It is contended that the making visible of textures of abandonment in the ways of working in mathematics can contribute towards enhancing the current teaching strategies teachers use in their mathematics classrooms. By using $\frac{(5a)^{-2}}{5a^{-2}} = \frac{(5a)^{-2}.a^2}{5} = \frac{5a}{5} = a$, John Mason suggests that learners should be accorded the opportunity to analyse the way of working to ascertain whether the simplification is correct. He asserts that exposing learners to "confusion [committing common errors], with the expectation that [learners] have been awakened to [them] … will make them more alert in the future" (Mason, 2000:10).

According to the results above, learners' 'doing and struggling' is no different from the way of working of adept mathematicians. Schoenfield (1987) suggests that learners should be exposed to such mathematics practices. In discussions with a colleague, this notion of exposing learners to confusion was not positively accepted because of the rigid focus on the correct thing. It is thus acknowledged that mathematician-like practices in the mathematics classroom are something to get used to and it should be seen as an attempt to move the practice of educators away from focusing on right and wrong and to rather attend to feedback, which draws attention to ways of working. However, to expect teachers to provide the kind of feedback linked to the analysis to bring forth the textures of ways of working is a daunting task. Teachers simply do not have the time to do this kind of analysis due to their other accountabilities. For this reason, it is contended that CPD providers should do this kind of analysis using the actual responses of learners' scripts produced from high-stakes examination contexts. These are then presented to teachers during CPD activities. During these interactions, teachers can contribute towards the creation of descriptors for the ways of working as was done with abandonment in order to build a repertoire of classroom-friendly constructs to describe the ways of working.

Exposing teachers to this kind of analysis will allow them to put more thought into the building of what will eventually become part of the arsenal of the candidates' historicised mathematical knowledge. Hence, the awareness of the little reminders in solution-seeking pursuance, which are exemplified in such analysis, will allow teachers to have some influence in building the historicised knowledge that will make sense and will be beneficial in a course of applying these little reminders in an examination context.

References

Coulon, A. 1995. *Ethnomethodology*. Thousand Oak: Sage Publications. https://doi.org/10.4135/9781412984126

Collins, H.M. & Yearley, S. 1992. Epistemological Chicken. In: A. Pickering (ed.), *Science as Practice and Culture*. Chicago: University of Chicago Press. pp.301-324.

Davis, R.B. 1984. *Learning mathematics: The cognitive science approach to mathematics education*. Norwood, NJ: Ablex Publishing Corporation.

DBE (Department of Basic Education). 2011. *Curriculum and assessment policy statement Grades 10-12*. Pretoria: DBE.

DBE (Department of Basic Education). 2014a. *National senior certificate: Grade 12, Mathematics P1, November 2014*. Pretoria: DBE.

DBE (Department of Basic Education). 2014b. *National senior certificate: Grade 12, Mathematics P1, November 2014*, memorandum. Pretoria: DBE.

DBE (Department of Basic Education). 2014c. *National senior certificate examination, 2014: Diagnostic report*. Pretoria: DBE.

Dufour, P. 2013. Practices of local social forums: The building of tactical and cultural collective action repertoires. In: P.G. Coy (ed.), *Research in social movements, conflicts and change* (Vol. 36). London: Emerald Group Publishing. pp.235-266. https://doi.org/10.1108/S0163-786X(2013)0000036011

Durkheim, E. 1938. *The rules of sociological method*. Translated by G. Catlin. (Original Work published 1895. Translated by S.A. Solovay and John H. Mueller. Edited by G. Catlin). New York: Free Press.

Garfinkel, H. 1967. *Studies in ethnomethodology*. Englewood Cliffs, NJ: Prentice Hall.

Garfinkel, H. 1991. Respecification: evidence for locally produced, naturally accountable phenomena of order, logic, reason, meaning, method, etc. in and as of the essential haecceity of immortal ordinary society (I) — an announcement of studies. In: G. Button (ed.), *Ethnomethodology and the human sciences*. Cambridge, MA: Cambridge University Press. pp.10-19.

Garfinkel, H. 2002. *Ethnomethodology's Program: Working Out Durkheim's Aphorism.* Lanham, MD: Rowman & Littlefield. https://doi.org/10.1017/CBO9780511611827.003

Greiffenhagen, C. 2008. Video analysis of mathematical practice? Different attempts to 'open up' mathematics for sociological investigation. *Forum Qualitative Social research/ Sozialforschung,* 9(3), Art. 32. Available: http://bit.ly/2ZzPLFz (Accessed 28 July 2019).

Holland, D.; Lachicotte Jr, W.; Skinner, D. & Cain, C. 2001. *Identity and agency in cultural worlds.* Cambridge: Harvard University Press.

Julie, C. 2003. Work moments in mathematical modelling by practicing mathematics teachers with no prior experience of mathematical modelling and applications. *New Zealand Journal of Mathematics,* 32:117-124.

Julie, C. 2015. Learners' dealing with a financial applications-;like problem in a high-stakes school-leaving mathematics examination. In: G. Stillman, W. Blum & M. Salett Biembengut (eds.), *Mathematical modelling in education research and practice: Cultural, social and cognitive influences.* New York: Springer. pp.477-486. https://doi.org/10.1007/978-3-319-18272-8_40

Liberman, K. 2012. Semantic drift in conversations. *Human Studies,* 35:263-277. https://doi.org/10.1007/s10746-012-9225-1

Livingston, E. 1986. *The ethnomethodological foundations of mathematics.* London: Routledge & Kegan Paul.

Livingston, E. 1987. *Making sense of ethnomethodology.* London: Routledge & Kegan Paul.

Livingston, E. 2006. Ethnomethodological studies of mediated interaction and mundane expertise. *The Sociological Review,* 54(3):405-425. https://doi.org/10.1111/j.1467-954X.2006.00623.x

Marick, B. 2004. Agile methods and agile testing. *STQE Magazine,* 3(5).

Mason, J. 2000. Asking mathematical questions mathematically. *International journal of Mathematical Education in Science and Technology,* 31(1):97-111. https://doi.org/10.1080/002073900287426

Matz, M. (1980). Towards a computational theory of algebraic competence. *The Journal of Mathematical Behavior,* 3(1):93-166.

Merz, M. & Knorr Cetina, K. 1997. Deconstruction in a 'Thinking' Science: Theoretical Physicists at Work. *Social Studies of Science,* 27(1):73-111. https://doi.org/10.1177/030631297027001004

Mugo, F.W. (2002). *Sampling in research.* Available: http://bit.ly/2IoBhK7 (Accessed 29 April 2017).

Pickering, A. 1993. The Mangle of Practice: Agency and Emergence in the Sociology of Science. *American Journal of Sociology,* 99:559-589. https://doi.org/10.1086/230316

Pickering, A. 1995. *The mangle of practice time, agency, and science.* Chicago, IL: University of Chicago Press.

Roth, W.M. 2013. *What more in/for science education: An ethnomethodological perspective.* Rotterdam/Boston/Taipei: Sense Publishers. https://doi.org/10.7208/chicago/9780226668253.001.0001

Schoenfeld, A. 1987. *Cognitive Science and Mathematics Education.* Hillsdale , NJ: Erlbaum Associates.

ten Have. 2004. *Understanding Qualitative Research and Ethnomethodology.* Sage Publications. Thousand Oaks: Sage Publications. https://doi.org/10.4135/9780857020192

| 10 |

EXAMINATION-DRIVEN TEACHING AS AN UNDERPINNING OF LEDIMATALI

Cyril Julie, Duncan Mhakure & Onyumbe Okitowamba

Introduction

Any CPD initiative has an objective. Dunst, Bruder & Hamby (2015: 1741) found in their meta-analysis of research reviews on CPD provision that "professional development may be more effective when it includes specific articulated objectives" Thus, for example, in a recent the multi-country European Union supported CPD project, PRIMAS – Promoting inquiry-based learning (IBL) in mathematics and science across Europe – the CPD objective is clearly stated as supporting teachers "by providing teaching materials along with professional development courses and a continuous support system within 'communities of IBL-practice'" (Maaß, Reitz-Koncebovski & Billy, 2013:ii).

One of the objectives for the Mathematics Teaching and Learning Initiative (LEDIMTALI) was set out in the funding call document. It is stated as "To improve the mathematics results (pass rates and quality of passes) in previously disadvantaged secondary schools as a result of quality teaching and learning" (National Research Foundation, NRF, 2011:3). The call was specific that the improvement in mathematics results should be in the

NSC Mathematics examination and by implication that the aim of continuing professional development should foreground this goal. Clearly then, CPD should focus on high-stakes examinations. To realise this goal, the LEDIMTALI project decided to make examination-driven teaching an underpinning focus of its CPD activities. More specifically, the issue pursued was the provision of professional development to high-school mathematics teachers to improve the quality of teaching for the enhancement of pass rates and the quality of passes in the NSC Mathematics examination. The next section focuses on high-stakes examinations.

High-stakes examinations

High-stakes examinations are defined in terms of the purposes they are used for. In line with the goals of LEDIMTALI as espoused previously, high-stakes examinations are viewed as school or externally set and marked national examinations which have consequences for candidates (Cizek, 2001; Yu & Suen, 2005). This means that upon being successful in the examination, the candidates obtain some credit, which allows them access to a variety of benefits. At the most elementary level within the schooling situation, it allows learners to progress from one grade to another. At the highest level of the schooling situation, it provides the candidates access to various post-school opportunities, such as job placements and enrolling for further education in tertiary institutions.

High-stakes examinations have a long history dating back to as early as 160 BC when the Chinese used them to select government officials based on how well individuals performed in the examinations (Gregory & Clarke, 2003; Pong & Chow, 2002). Since then, there has been a growing use of the high-stakes examinations and an increasingly diverse number of types of such examinations are being used in different educational systems globally. In line with the definition in the former paragraph, high-stakes examinations serve several functions in a country. These include:

Ø   as a norm-referenced assessment for selecting students who have the required academic competencies to attend higher education institutions;

Ø   certifying mastery of skills to gain certification as an artisan; and

Ø   levelling the testing field, thus eliminating patronage and corruption in the examination system (Boardman & Woodruff, 2004; Chapman & Snyder, 2000; Heyneman & Ransom, 1990; Wall, 2000).

Lorrie A. Shepard acknowledges that in the majority of cases it was the teachers who, to ensure fairness, "believed that assessments had to be uniformly administered, so they were reluctant to conduct more individualised [teacher-set] assessments" (Shepard, 2000:5). 'Fairness' in this context can also be regarded as a way of eliminating potential testing inequalities that could be used to preserve certain class-based elitism. This 'fairness' is believed to lead to the democratisation of the education system and so leading to "making education a possibility for all" (Amit & Fried, 2002:500). Amit and Fried also posit that high-stakes testing in some countries is construed as the "product of a nationally minded, achievement-oriented pedagogy", which is aimed at, among other goals, promoting key imperatives, such as mathematical literacy and numeracy (Amit & Fried, 2002:500).

High-stakes examinations as a form of summative assessment in mathematics are not an uncontested area but proponents have argued that they have "become a permanent and vital part of education" (Wideen, O'Shea, Pye & Ivany, 1997:430). Despite the variety of views on high-stakes examinations, there seems to be a convergence that it is necessary but there is a need for a wider set of 'competencies' to be included in line with current suggested reforms for school mathematics. In fact, for the realisation of these reforms, it is suggested that high-stakes examinations be used as leverage for reform initiatives due to these examinations being the translator of the espoused curriculum into the implemented curriculum (Au, 2007). From the reform perspective, a criticism levelled against time-restricted single-sitting high-stakes examinations is that they "often fail to address important process goals for mathematics, such as the ability to engage in reasoning and proof, communicate about mathematics, solve non-routine problems, or represent mathematical concepts in multiple ways" (Suurtamm et al., 2016:21). Although there is truth in this statement, there are, at least as far as the South African NSC Mathematics examination is concerned, items (or sub-items of questions) which do incorporate some of the desired process goals. For example, the following item (Question 3) requires the form of process reasoning skill referred to by Watson and Mason (1998:14) as "comparing, sorting and organizing" epitomised by the phrase "What is the same and what is different about …"

QUESTION 3

Consider the series: $S_n = -3 + 5 + 13 + 21 + \ldots$ to n terms.

3.1   Determine the general term of the series in the form $T_k = bk = c$

3.2   Write S_n in sigma notation.

3.3   Show that $S_n = 4n^2 - 7n$.

3.4   Another sequence is defined as:

$Q_1 = -6$

$Q_2 = -6 - 3$

$Q_3 = -6 - 3 + 5$

$Q_4 = -6 - 3 + 5 + 13$

$Q_5 = -6 - 3 + 5 + 13 + 21$

3.4.1   Write down a numerical expression for Q_6.

3.4.2   Calculate the <u>value</u> of Q_{129}.

FIGURE 10.1   An examination question in the high-stakes NSC Mathematics examination containing a process section

It can be observed that Q 3.1 to 3.3 are reasonably routine with different levels of cognitive demand. For 3.4 the "What is the same and what is different about …" comes into play as candidates have to identify that -6 has been added to the originally given series. The number and variety of process skills (or higher cognitive level demand) items included in high-stakes examinations are governed by the prescriptions given to examination designers in the curriculum documents on which the examination is based. It appears that even if examinations are entirely based on the reform ideals, there will always be a distribution of items according to cognitive demand to discriminate between learners performing at different levels. The 2016 DBE report on the examination draws attention to Q 3.4 by stating that "Candidates struggled to understand Q 3.4 and many did not see the connection between Q 3.3 and Q 3.4" and suggests that learners must be exposed to 'unseen' type questions where unfamiliar patterns are formed and convince them that "these are generally easy to solve" (DBE, 2016:155). The next section considers the notion of examination-driven teaching and high-stakes examinations documented in the research literature.

Examination-driven teaching

The assertion that "Assessment should reflect the mathematics that is important to learn and the mathematics that is valued" (Suurtamm et al., 2016:5), is an acknowledgement that valued and legitimate school mathematics knowledge is to a large extent determined by the assessed curriculum. This acknowledgement of the primacy of the structuring effect that the assessed curriculum has on the enacted curriculum is widely reported. Burkhardt (1987) refers to it as the WYTIWYG (What You Test Is What You Get) phenomenon, which is also the description given by Ruthven (1994). For the LEDIMTALI project, it is referred to as the What Is Tested Is What Is Taught (WITIWIT) phenomenon. WITIWIT is also equally true for high-stakes examinations because these examinations are guided by rigid curriculum guidelines, which define how candidates are tested. By defining how candidates are tested, there is an expectation of how they should be taught to prepare them to achieve the objectives of the high-stake examinations.

The structuring influence of high-stakes examinations had the effect that teaching is driven by the examinations and hence the notion of examination-driven teaching. Examination-driven teaching is normally viewed as "teaching the content of previous examinations and anticipated questions that might crop up in an upcoming examination of the subject" (Julie, 2013:1). Some other terms used for teaching influenced by assessments and learners' performances on such assessments are measurement-driven teaching, examination-driven instruction, measurement-driven instruction, assessment-driven instruction, data-driven instruction and data-driven decision-making. They differ in intent, but all use examinations and/or learners' performance as data sources to derive methods of teaching to improve achievement in such examinations.

A weak notion of examination-driven teaching is that of teaching to the test. The notion of teaching to the test is conceptualised as:

> Classroom practices that emphasise remediation, skills-based instruction over critical and conceptual oriented thinking, decreased use of rich curriculum materials, and narrowed teacher flexibility in instructional design and decision making, and the threat of sanctions for not meeting externally generated performance standards. (Davis & Martin, 2008:11)

This position, although it also alludes to other issues related to motives for the adoption of examination-driven teaching, ignores the fact that there can be items in high-stakes examinations of the "critical and conceptual oriented thinking" kind. In fact, with items of this nature "teachers who teach to the test [can] deliver a rich and balanced curriculum" (Swan & Burkhardt, 2012:5). Popham (1987), also views the focusing on what is expected in examinations as somewhat inevitable.

This is due to what the consequences of success or not in these examinations have for teachers, learners, educational authorities and parents. Further, it is argued that examination-driven teaching has important advantages, such as the clarity of instructional goals, cost-effectiveness for improving the quality of education and the provision of valuable feedback to teachers for instructional decision-making (Popham, 1987; Shepard & Dougherty, 1991; Wayman, 2005). Regarding its effects, examination-driven teaching is thus viewed as an opportunity and catalyst for the improvement of mathematics achievement (Burkhardt, 2006).

Cizek (2001:26) captures the dilemma of high-stakes examinations and its impact on teaching succinctly by asserting that "High-stakes tests: we do not know how to live with them; we cannot seem to live without them".

Despite the possible negative consequences of examination-driven teaching on education reform, studies by Bishop (1998, 2000) have given some positive appraisals of the effects of examination-driven teaching. Bishop (1998) compared the performance of students on the International Assessment of Educational Progress (IAEP) and Third International Mathematics and Science Study (TIMSS) of countries around the world and some Canadian provinces and the results indicate that there is a positive correlation between the high-stakes examination scores and the students' scores in IAEP and TIMSS. Bishop (2000) also demonstrated that students in states that were using high-stakes examinations as exit tests "performed significantly better on the National Assessment of Educational Progress (NAEP) in 8th grade, and on the Scholastic Assessment Test (SAT) in high school" (Cizek, 2001:26). The important point here is that Bishop (1998, 2000) has shown that high-stakes examinations can improve the student's learning as evidenced by the student's performance. This is contrary to the rhetoric about examination-driven teaching being a hindrance to the student's performance.

Another emerging outcome of examination-driven teaching is that it seems to address achievement gaps between different societal groups. Ashley and Hand allude to this and suggest that examination-driven teaching "seems to be accomplishing some good (i.e., increased proficiency in basic skills and a narrowing of the black-white achievement gap)" (Ashley & Hand, 2007:2). On the contrary, it is argued that the use of examination-driven teaching increases social inequalities (Yu & Suen, 2005). Yu and Suen's position is similar to that of Davis and Martin (2008), who are explicit that teaching strategies driven by focusing on making learners competent in developing low-level skills are part of the larger structuring effects of standardised tests to contribute towards African-American learners being in a position of subservience. As pointed out above, an examination-driven teaching strategy can be used in a more empowering fashion by including the development of the learners' competencies that deal with mathematical items involving "critical and conceptual oriented thinking" But as Davis and Martin (2008) posit, the teaching must be accompanied by a strong belief that learners from marginalised groupings can be doers of mathematics. Also, as Tate (1993) posits, performance in high-stakes mathematics examinations should be used as forms of feedback to foster learning and not to rank learners.

Although it is a truism that the achievement of marginalised groupings is measured by standards set by the non-marginalised, the assumption that teaching in these environments is not subjected to examination-driven approaches is not necessarily true. Teese for example, argues that:

> It is folly to close one's eyes to what is happening in schools for learners from high-stakes examinations environments…[where] classes can be worked to a high standard of competitive performance, following the bureaucratic procedures of syllabus-stripping (reduction to examinable topics), chalk and talk, worked examples, worksheets, past papers and exam rehearsals. (Teese, 2000:186)

This quotation vividly describes how the mathematics teaching space is organised in high-stakes examinations schools in the state of Victoria in Australia. Clearly, the focus is on getting the students to succeed in high-stakes examinations.

For LEDIMTALI, examination-driven teaching underlies its activities with high-stakes examinations taken as the demarcater of legitimate school mathematics knowledge. However, care is taken not to restrict teaching to low-level, skills-based instruction with meaningless drill, but rather to use examinations in a variety of ways to foster the learners' 'mathematicalnesss' to enhance their achievement in high-stakes examinations.

The operationalisation of examination-driven teaching in LEDIMTALI

In LEDIMTALI, examination-driven teaching is operationalised in various ways. Firstly, there was careful listening to the dilemmas teachers expressed to seek common solutions to resolve these dilemmas. These dilemmas were the normal issues related to shortcomings, dispositions and attitudes of learners. The project had to 'listen' to teachers but also not allow these issues to detract from the objective of the project. Issues beyond the control of teachers' teaching and learners' learning had to be suppressed.

A major issue raised by teachers was that "learners do not do homework". This led to discussions on homework and its purposes. One of the purposes of homework is consolidation and developing procedural fluency. The discussions culminated in the conclusion that a strategy needs to be devised to address the issue of forgetting. It was suggested that a strategy akin to one that was used in the 'old days' in primary schools be considered (see Chapter 4 of this book). This strategy entails that before the start of the day's lesson the teacher would give learners some oral problems, normally linked to the four basic operations, on previously done work. This discussion culminated in a strategy labelled 'spiral revision', where at the start of a lesson learners are presented with two or three problems or exercises dealing with previously taught work to solve. Teachers readily adopted this strategy, albeit that they adapted it to fit their circumstances. Some teachers would, for example, set aside an entire period per week for 'spiral revision'. A carefully controlled study on this strategy was conducted. The study focused on the quadratic theory section of the first examination paper for Grade 12 and the straight-line analytic geometry and manipulative trigonometry sections of the second paper. The results of the study indicate that the outcomes of the implementation of the strategy were positive in terms of Grade 12 learners improving their achievement scores on the mentioned topics from the June to the September examinations (Julie, Mbekwa & Simons, 2016). An unintended outcome of the use of the "spiral revision strategy" is related to learners' motivation. A learner who previously asserted "I am not going to focus on mathematics and put my energies in my other subjects' changed his view to 'I am beginning to see that I can make it in mathematics'".

As an aside, it is important to note that the 'spiral revision' strategy was home-grown and readily adopted by the participating teachers. As the project progressed, an associated construct developed was 'quickies' (see also Chapter 4), which refers to the activities that are used with the "spiral revision" strategy. Critical race theory in Mathematics Education (Ladson-Billing, 2000; Ladson-Billing & Tate, 1995; Martin, 2009) recommends that constructs coined by teachers, regardless of whether they are part of the dominant academic discourses in teaching, be valued and adopted to counter the notion of a 'pedagogy of poverty' (Haberman, 1991:45). This, it can be argued, dominates many CPD initiatives in South Africa for African and South African First Nation and Slave Descendant teachers with a structuring notion of "working with teachers from where they are" at the expense of "working with teachers with what they bring and can be developed".

Figure 10.2 gives a set of 'quickies' developed by teachers during an institute to address specific problems learners experienced and those highlighted in the diagnostic reports on the NSC Examination (DBE, 2016). For this particular quickie, the DBE report states: "The original function must be in the differentiable form (that is, in terms where one can correctly identify the coefficient, variable and exponent) before the rules of differentiation can be applied" (DBE, 2016:159).

GRADE 12: QUICKIES SET 1 – Calculus

Writing an expression in the form of separate terms and each term not having the variable as a denominator.

Write the following expressions as one with separate terms, the variable not as a denomiator and, where applicable, not root signs in the expression.

A. $\frac{x+5}{\sqrt{x}}$ (Ans:$x^{\frac{1}{2}} + 5x^{-\frac{1}{2}}$)

B. $\frac{15x^2+x-2}{3x-1}$ (Ans: $5x + 2$)

FIGURE 10.2 Example of a 'quickie' developed to deal with an identified issue in the calculus section of the NSC examination

The teachers developed activities of this nature to address a particular problem in that learners are used to simplification of algebraic fractions where the pursuit is normally to obtain a single expression in earlier grades. Little attention is given to reverse of the simplification process – deconstructing a simplified algebraic fraction into constituent parts. The former process, simplification was given the label 'assembly' and the latter one, deconstruction, 'disassembly'. This is again an instance of the project leaders and the participating teachers developing labels or language terms to allow them to have conversations about these matters.

A second way that focus on examinations was operationalised is what we termed "designing (or mapping) down" to lower grades the mathematics content in the final high-stakes NSC Mathematics examination. This way of designing learning activities arose from early observations of classroom lessons where lesson topics are dealt with in a compartmentalised manner and confined to a particular grade with little connection to the competencies that are required to be successful in the high-stakes NSC Mathematics examination. For example, in Grade 10, with respect to the system of real numbers, the National Curriculum and Assessment Policy Statement (CAPS) states, amongst other things that learners must "Understand that real numbers can be rational or irrational" (DBE, 2011:21) This flows from the curriculum of previous Grades 8 and 9. The knowledge regarding the real number system must be consolidated in Grade 10.

From time to time, LEDIMTALI accesses the actual NSC Mathematics examination scripts of the participating schools from the Western Cape Department of Education. These scripts are analysed with the view of feeding back to teachers the nature of the ways candidates dealt with the examination items. Figure 10.3, obtained through a Rasch analysis, gives a sense of the difficulty levels of the items for the NSC Mathematics examination (Paper 1) for 2012.

FIGURE 10.3   Rasch analysis of the first paper of the NSC Mathematics examination of the project schools.

Table 10.1 gives the classical item difficulty for the items which candidates found the most difficult.

TABLE 10.1   Classical item difficulty

Item	1.3.2	10.2	4.1.4	3.3.2	9.2.2	5.1	11.3.3
Item difficulty	0,01	0	0,02	0,05	0,01	0,01	0

Item 1.3.2, which carried 1 mark, reads as follows:

The solutions of a quadratic equation are given by $x = \dfrac{-2 \pm \sqrt{2p+5}}{7}$
For which value(s) of p will this equation have no real solutions?

In workshop discussions, teachers found it surprising that the item was found so difficult. According to them, this is knowledge that learners deal with from Grade 10 onwards.

The spill over of this kind of engagement was that teachers planned to incorporate NSC Mathematics-like problems in the repertoire of problems to deal with in Grade 10. During the one institute, planning for the first week of teaching was done for teaching the next year (see Table 10.2).

The near similarity of the expression $\sqrt{\dfrac{9}{11-x}}$ with $x = \dfrac{-2 \pm \sqrt{2p+5}}{7}$ is observable and exemplifies the 'design down' idea since the nature of roots is only dealt with in Grade 11. However, in dealing with the real number system in Grade 10, this topic is already dealt with.

TABLE 10.2   Grade 10 planning for the first week of 2013

Week	Day 1	Day 2	Day 3	Day 4	Day 5
1	Baseline assessment (from Grade 9). Introduce: Concepts, e.g. real, integer, rational, irrational, etc. Is 2 a rational number? Why is x the multiplicative inverse of Y? What happens when you add 0 to a real number? What happens when you multiply a real number by 1? Use knowledge level to apply concepts N / N₀ / Z / Q₊ / Q⁻ / R / √2	Non-real and undefined. Simple procedures to identify rational, irrational. E.g. Consider: $A = \sqrt{\dfrac{9}{11-x}}$ For which values of x will A be rational, irrational, undefined, non-real? Homework	Complex procedures. E.g. Determine without a calculator between which consecutive integers $\sqrt{51}$ lies.	Problem-solving. E.g. Changing recurring decimals to rational. Prove that $0,\dot{2}4\dot{5}$ is rational. Show that $2 + \sqrt{3}$ is an irrational number.	Assessment Tutorial Test

The idea of designing down from the high-stakes NSC Mathematics examination gained traction and was later included in a project-designed module for the real number system for Grade 10. In addition to having activities to develop procedural fluency, the module also contains questions to deepen thinking of the evaluative kind, as indicated in Figure 10.4.

Sipho said he discovered a way to determine whether a number written as a fraction with a square root of an algebraic expression is a real, rational number. He explained his method as below with his example as $A = \frac{-2 - \sqrt{b-3}}{3}$. Steps Sipho followed.	Result of the step
1 Write down the expression under the square root (√)-sign.	$b - 3$
2 Find b by adding the 3 to any square number. (Sipho chose 25)	$b = 25 + 3 = 28$
3 Write the number A with the value of b found in step 2.	$A = \frac{-2 - \sqrt{28-3}}{3}$
4 Simplify the expression under the square root (√) sign.	$A = \frac{-2 - \sqrt{25}}{3}$
5 Simplify the numerator.	$-2 - \sqrt{25} = -2 - 5 = -7$
6 Write the number with the numerator found in step 5.	$A = \frac{-7}{3}$
7 The number found in step 6 is a real, rational number.	$\frac{-7}{3}$ is a real, rational number.

FIGURE 10.4 An evaluative activity linked to high-stakes NSC Mathematics examination type questions for Grade 10

This section provided a glimpse on how the knowledge that is privileged in the ultimate high-stakes examinations is filtered down to lower grades to allow learners to engage with this knowledge, albeit in a more elementary fashion. Grade 12 teachers appreciate the value of this approach. It is seen as a way of viewing school mathematics as coherent across all grades in high school (Grades 8-12) with the objective of getting all teachers – be they in-field (qualified to teach school mathematics as primary teaching subject) or out-of-field (qualified to teach school subjects other than mathematics as primary subject) teachers – having the singular goal of enhancing achievement in the ultimate high-stakes NSC Mathematics examination, regardless of the grade level they are teaching

Where feasible, activities for Grade 8 are considered. For example, in recent discussions, the idea was suggested that for simplification of algebraic fractions, a scaffolded activity (as presented in Figure 10.5) should be dealt with in Grade 8. This would help to develop procedural fluency for dealing with the determination of the derivative, using the formula, from first principles in the high-stakes NSC Mathematics examination.

The expression A = 2x is given.

1. Form a new expression B where x is replaced with $x + h$.
2. Determine B − A
3. Now calculate $\frac{B - A}{h}$

FIGURE 10.5   An example of a designed-down activity for Grade 8

Learner performance in the NSC Mathematics high-stakes examination

Three measures for learner performance were highlighted for improvement in the concept document for the Research and Development chairs in Mathematics Education. These are an increase in (a) the number of learners enrolling for the high-stakes NSC Mathematics examination, (b) the pass rate of the learners, and (c) the quality of the passes. The latter category comprises two sub-categories, namely the average percentage scores and the levels of performance. Any assessment of learner performance over some time needs, at a minimum, methods of analysis of the performances and theories/hypotheses or credible established research outcomes to interpret the outcomes of the analysis. The latter is dealt with under a commentary on the performance outcomes over the four years.

There are generally three elementary models to assess improvement in performance. There are more sophisticated models but these require access to the entire set of raw data of individual candidates for analysis (Van Geel, Keuning, Visscher & Fox, 2016). One elementary model is the absolute headcount decrease/increase model. Its importance is normally underestimated, but its strength is that it gives a good idea on whether sufficient people are supplied for a system. The model is suitable if, for example, the system requires X persons with a school-leaving certificate to fulfil its requirements to sustain a sufficient supply of qualified staff. Another elementary model is the year-on-year model which compares the average performance scores on a year-on-year basis. The last elementary model is the baseline comparative mode, which compares the average performance scores relative to an average score of a baseline. All three models have their advantages and disadvantages. Conventionally, the latter one is preferred to measure progress or not since it has some sensitivity to changing conditions by having a fixed reference point.

From an educational measurement perspective, the measure of importance is the average for the period for the measurements. The reporting on measures that follows is done for the three elementary models. For the baseline comparative model, the 2011 measures are taken as baseline measures. Ten schools were approved by the Western Cape Education Department to participate in the project. One of these schools was dropped due to unsatisfactory attendance of the project activities. The analysis gives the outcomes for the nine schools.

Learners who wrote the NSC Mathematics examination for the period 2012-2015

TABLE 10.3   Enrolments, 2012-2015

Model		2011 (Baseline year)	2012	2013	2014	2015
Headcount	Number	256	248	358	375	422
	Year-on-year increase		-8	102	17	47
Year-on-year	% increase		-3%	44%	5%	13%
	Average increase					15%
2011 baseline	Number		-8	102	119	166
	Increase from baseline year		-3%	40%	40%	65%
	Average increase					37%

The headcount model indicates that the stipulated 10% increase in enrolments was surpassed. The other two models show that the number of learners who wrote the high-stakes NSC Mathematics examination increased by an average of at least 15% for both the year-on-year and the baseline models.

Pass rates

TABLE 10.4 Pass rates, 2012-2015

Criteria	Description	2011	2012	2013	2014	2015
Model	Total number of learners	256	248	358	375	422
Headcount	Number of learners passed	81	129	142	142	163
	Increase		48	61	61	82
Year-on-year	Pass rate (%)	32%	52%	40%	38%	39%
	Increase (%)		20%	-12%	-2%	1%
	Average increase (%)					2%
2011 baseline	Increase (%)		20%	8%	6%	7%
	Average increase (%)					10%

All models indicate that on average the pass rates did not decrease. At worst it increased on average by 2% for the year-on-year model.

Average percentage mark

TABLE 10.5 Average pass percentage mark, 2012-2015

Criteria	Description	2011	2012	2013	2014	2015
Model	Average scores (%)	28%	35%	28%	26%	29%
Year-on-year	Increase (%)		8%	-7%	-2%	3%
	Average increase (%)					0,4%
2011 baseline	Increase (%)		7%	0%	-2%	1%
	Average increase (%)					2%

The average percentage marks did not decrease. It increased marginally (0.4%) for the year-on-year model and increased by 2% for the baseline-comparative model.

Levels of performance

The levels of performance shown in Tables 10.6 and 10.7 are those that are given in the CAPS document (DBE, 2011:56). These levels of performance are accompanied by percentages of achievement bands. The analysis used these seven percentage achievement bands. Table 10.6 shows an analysis of students' levels of achievements between 2011 and 2015.

Table 10.6   Levels of performance, year-on-year, 2012-2015

Rating codes (Levels of performance (%)	2011-2012	2012-2013	2013-2014	2014-2015	Average increase	Direction of shift
1 (0-29%)	-18,81%	12,35%	1,80%	-0,76%	-1,36%	–
2 (30-39%)	9,39%	-3,06%	1,39%	-3,12%	1,15%	+
3 (40-49%)	5,13%	-7,69%	0,78%	2,28%	0,12%	+
4 (50-59%)	3,78%	-1,21%	-3,53%	1,48%	1,895%	+
5 (60-69%)	1,66%	0,13%	-2,02%	0,56%	1,09%	+
6 (70-79%)	-0.38%	-0,12%	1,05%	-0,15%	0,10%	+
7 (80-100%)	-0,77%	-0,40%	0,53%	-0,30%	-0,24%	–
Summary of level shifts	Five shifts in the positive direction and two in the negative direction. The negative shift in 0-29% band is a positive since it indicates a decrease in the number of failures.					–

Table 10.7   Levels of performance, baseline, 2012-2015

Rating code (Levels of performance) (%)	2011 Baseline					Average increase	Direction of shift
	2011	2012	2013	2014	2015		
1 (0-29%)	66,80%	47,98%	60,34%	62,13%	61,37%	-8,84%	–
2 (30-39%)	16,02%	25,40%	22,35%	23,73%	20,62%	7,02%	+
3 (40-49%)	8,98%	14,11%	6,42%	7,20%	9,48%	0,30%	+
4 (50-59%)	4,69%	8,47%	7,26%	3,73%	5,21%	1,17%	+
5 (60-69%)	1,56%	3,23%	3,35%	1,33%	1,90%	0,85%	+
6 (70-79%)	0,78%	0,40%	0,28%	1,33%	1,18%	0,00%	Unchanged
7 (80-100%)	1,17%	0,40%	0,00%	0,53%	0,24%	-0,91%	–
Summary of level shifts	Four shifts are in the positive direction, two in the negative direction and one unchanged. The negative shift in the 0-29% band is a positive since it indicates a decrease in the number of failures.						

An analysis of both models indicates that there is only one unwelcome negative shift in the rating code 7.

An inspection of the number of learners who wrote the NSC Mathematics examination for the period 2012 to 2015 indicates that the number of learners increased from 248 in 2012 to 422 in 2015. The increase in the number of learners per school varied. One school, for example, had five learners who wrote the NSC Mathematics examination in 2012 and the school obtained a 100% pass rate. In 2013, the number of learners increased to 48 and the pass rate decreased to 8%. In another school, the number of learners increased from 42 in 2012 to 69 in 2014. The pass rate, however, remained below 10%.

To further investigate the improvement in learner achievement in the NCS Mathematics examination, the results of the project schools were compared to a stratified sample of nine schools. The stratification was according to school district, circuit and the project team's assessment of similarity of socio-economic conditions. The following graphs present a picture of this comparison.

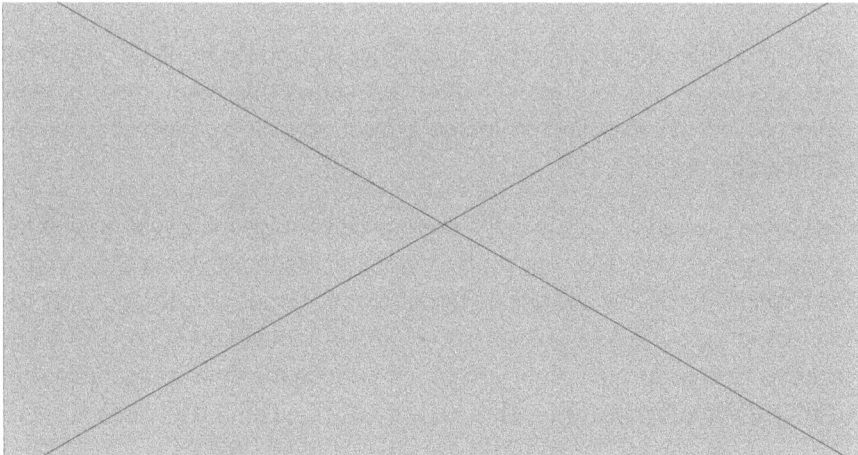

FIGURE 10.6   Comparison of learner enrolment between project schools and a set of comparable schools

Figure 10.6 illustrates that for the absolute headcount measure the increase of examinees who wrote the NSC Mathematics examination was more pronounced for the project schools.

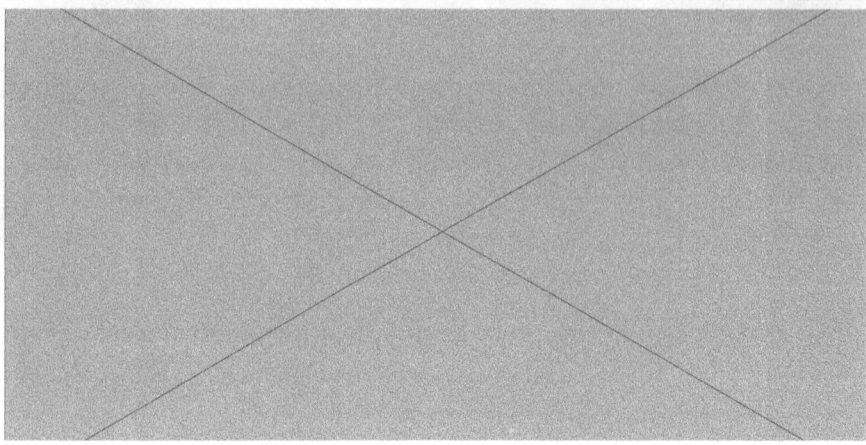

FIGURE 10.7   Comparison of number learners who passed of project schools and a set of comparable schools

As is the case with the number of learners who wrote the NSC Mathematics examination for the four years, Figure 10.7 shows that the number of learners who passed the examination in the project schools surpassed those in the comparable schools.

The overall average increase in the average percentage mark year-on-year for the project schools was 0,552 and for the comparable schools it was -0,01. With 2011 as baseline the overall average increase for the average percentage mark for the project schools was 2,08 whilst it is -0,66 for the comparable schools. The project schools thus performed slightly better on average than the comparable schools with respect to the average percentage mark obtained (Effect sizes (Cohen's d: 0,21 year-on-year and 0,28 on 2011 baseline)) (Cohen, 1988).

The year-on-year average pass rate increase for the period was 2% for the project schools and 0% for the comparable schools with an effect size on the difference as 0,31. For the 2011 baseline, the average pass rate increases are 9% and -8% for the project and comparable schools respectively with an effect size of 0,71. This indicates that on average the increase in the pass rate was above moderate for the project schools relative to the comparable schools.

Regarding the levels of achievement, the overall average shows an increasing trend on all the levels of achievement bands except the highest and the lowest band whilst the trend for comparable schools was an increase in the band 0-29% (the failing band), 30-39%, 70-79% and 80-100%. For the other bands, the comparable schools show a decreasing trend. Figure 10.8 illustrates this situation.

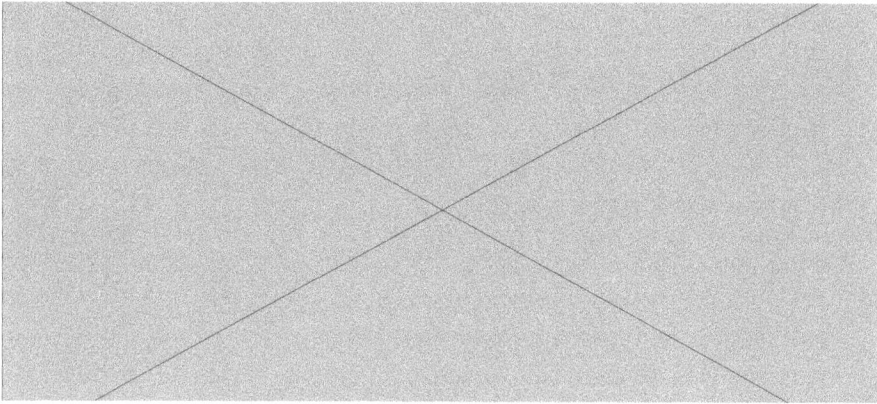

FIGURE 10.8   Comparison of average levels of performance of project schools and a set of
comparable schools

Conclusion

This chapter focused on the issue of whether the underpinning examination-driven strategy of LEDIMTALI has the potential for enhancing achievement outcomes in the high-stakes NCS Mathematics examination. The results referred to in the chapter indicate that there is an overall movement from "low performing schools" to "previously low performing schools". In other words, schools that were referred to as "low performing schools" have improved in their mathematics performance such that they are now known as "previously low performing schools".

This movement is not pronounced and raises various questions. One question revolves around "What accounts for these non-pronounced shifts in achievement performance of learners from low socio-economic environments?" The best recourse, in our opinion, is to research literature on change in achievement in high-stakes mathematics examination, theories and hypotheses related to change in social phenomena. Change in social phenomena is normally regarded as transformation. According to the critical realist approach to social phenomena, transformation is by nature slow, as demonstrated by Bhaskar's Transformation Model of Social Activity (Archer, Bhaskar, Collier, Lawson & Norrie, 2013). It can be argued that any effort related to change in scholastic achievement is a transformation activity related to inequality due to achievement being structured along socio-economic dimensions. Archer, et al. (2013) argue that at the initial stages of a transformation initiative change is so miniscule that it is hardly observable. The aforementioned results are in accordance with this theory. Another plausible explanation is that there is a tension between the two goals of the project, namely increasing the quality of teaching, and increasing the number of students studying mathematics, and thus including weaker students to study mathematics. In this study, the inclusion of more students including weak students could have caused marginal gains in the students' achievements in high-stakes examinations. Therefore, it is recommended that future similar studies should maintain a fixed number of students taking mathematics throughout the duration of the study.

Moving into research results directly related to achievement in school mathematics, a comprehensive study was conducted by the RAND Corporation in the United States on the closing of the "achievement gap in mathematics" between Caucasian, black and Latino 17-year olds (Berends, Samuel, Sullivan & Briggs, 2005). The sample of students involved covering the 20 years was a nationally representative one comprising 14 469 students from 875 schools. The data used were, amongst other instruments, obtained from the National Assessment of Educational Progress (NAEP) mathematics tests. The RAND Corporation study found that the achievement gap in mathematics between Caucasian and black students decreased from an average standard deviation of 1,14 to 0,68 over 20 years. In about the same period, the Caucasian-black achievement gap decreased on average by $\frac{1}{100}$ th of a standard deviation per year with fluctuations of increases and decreases over the period. For the Caucasian-Latino situation, the average difference in standard deviations dropped from 0,94 to 0,66. The closure of the achievement gaps resulted from massive investment in educational and social upliftment programmes in the USA over the mentioned period. The results indicate the slow progression of improvement in learning gains as measured by a widely used mathematics national examination in the USA. The Berends et al. (2005)'s findings though similar to the LEDIMTALI, need to be interpreted with caution because the contexts are different. In the same vein, as a limitation, the dynamics in terms of numbers of students taking mathematics in comparative schools might not have increased with the same proportions as in the LEDIMTALI schools – these could well have influenced the findings.

Given the slow but positive progression of improvement in learning in the LEDIMTALI project, the issue of teachers sustaining the implementation of the tactics and strategies that contribute towards enhanced achievement is an important one. This is linked to the so-called 'dosage' time of the CPD. Cordingley, Bell, Rundell, Evans and Curtis (2003:64, italics in original) suggest, based on meta-synthesis of research results of collaborative CPD, that "*arrangements* [be made] *for sustaining learning over time so that new approaches can be adapted, experimented with and integrated incrementally into existing practices*".This suggestion points in the direction of extended engagement in

CPD to counter fall-back into pedagogical procedures which have less potential to realise the enhanced achievement goal. Later meta-synthesis studies (Dunst et al., 2015; Walter & Briggs, 2012) concur with this result related to dosage time. The dosage time refers to the actual time that teachers are involved in and to a certain extent the number of years the teachers were supported – the total project time. The meta-synthesis studies do not provide information on the overall number of years for the strategies and tactics to become fully part of the repertoire of strategies and tactics teachers employ to enhance achievement. Carpenter, Fennema, Loef Franke, Levi and Empson (2000:4) report on a three-year CPD program involving 21 teachers. They "explicitly examined the nature and pattern of change among teachers and the relation between beliefs and instruction". They used a four-level scheme of teacher classroom engagement and found that "19 of the 21 teachers in the longitudinal study were at Level 3 or higher … Eighteen of the 21 teachers had changed at least one level in beliefs and practice, and twelve had changed at least two levels" at the end of the CPD-initiative (Carpenter et al., 2000:5). It appears reasonable to make the inference that a dosage time of three years can improve achievement outcomes.

Carpenter et al. (2000) did a follow-up study four years after the CPD initiative had ended. This study showed that "all of the teachers continued to implement principles of the programme at some level. Five of the teachers had slipped one level, but ten of the teachers showed continued growth" (Carpenter et al., 2000: 5). The results suggest that a minimum of three years of sustained and collaborative CPD can bring about positive shifts in teaching aimed at a particular objective. The three years is obviously context-bound and for the studies mentioned above, the focus was on primary school classrooms. The meta-synthesis, however, brings to the fore the need for "follow-up supports … after the completion of the initial in-service professional development" (Carl, Duns, Bruder & Hamby, 2015:1738), to ensure embeddedness of examination-driven teaching, as conceptualised in LEDIMTALI, to increase achievement in high-stakes examinations in a more pronounced way.

References

Amit, M. & Fried, M.N. 2002. High-stakes assessment as a tool for promoting mathematical literacy and the democratisation of mathematics education. *Journal of Mathematical Behaviour*, 21:499-514. https://doi.org/10.1016/S0732-3123(02)00147-5

Archer, M.; Bhaskar, R.; Collier, A.; Lawson, T. & Norrie, A. (eds.). 2013. Critical realism: Essential readings. Abingdon-on-Thames, UK: Routledge.

Ashley, T. & Hand, E. 2007. Reflection paper on measurement-driven instruction. Available: http://bit.ly/2lt5mbD (Accessed 28 July 2019).

Au, W. 2007. High-stakes testing and curricular control: A qualitative metasynthesis. *Educational researcher*, 36(5):258-267. https://doi.org/10.3102/0013189X07306523

Berends, M., Samuel, R L., Sullivan, T. & Briggs, RJ. 2005. *Examining gaps in mathematics achievement among racial ethnic groups, 1972-1992*. Santa Monica, CA: RAND Corporation. https://doi.org/10.7249/MG255

Bishop, J.H. 1998. The effect of curriculum-based external exit exam systems on student achievement. *The Journal of Economic Education*, 29(2):171-182. https://doi.org/10.1080/00220489809597951

Bishop, J.H. 2000. Curriculum-based external exit exam systems: Do students learn more? How? *Psychology, Public Policy, and Law*, 6(1):199-215. https://doi.org/10.1037//1076-8971.6.1.199

Boardman, A.G. & Woodruff, A.L. 2004. Teacher change and 'high-stakes' assessments: What happens to professional development? *Teaching and Teacher Education*, 20(6):545-557. https://doi.org/10.1016/j.tate.2004.06.001

Burkhardt, H. 1987. Curricula for active mathematics. In: I. Wirszup & R. Streit, (eds.), *Developments in school mathematics around the world* (Vol.1). Reston, VA: National Council of Teachers of Mathematics. pp.321-361.

Burkhardt, H. 2006. Modelling in mathematics classrooms: Reflections on past developments and the future. *ZDM*, 3(82):178-195. https://doi.org/10.1007/BF02655888

Carl, J.; Duns, C.J.; Bruder, M.B. & Hamby, D.W. 2015. Metasynthesis of in-service professional development research: Features associated with positive educator and student outcomes. *Educational Research and Reviews*, 10(12):1731-1744. https://doi.org/10.5897/ERR2015.2306

Carpenter, T.P.; Fennema, E.; Loef Franke, M.; Levi, L. & Empson, S.B. 2000. *Cognitively guided instruction: A research-based teacher professional development program for elementary school mathematics*. Madison, WI: The National Center for Improving Student Learning & Achievement (NCISLA) in Mathematics & Science, University of Wisconsin-Madison.

Chapman, D.W. & Snyder, C.W. 2000. Can high stakes testing improve instruction: Re-examining conventional wisdom? *International Journal of Educational Development*, 20(6):457-474. https://doi.org/10.1016/S0738-0593(00)00020-1

Cizek, G.J. 2001. More unintended consequences of high-stakes testing. *Educational Measurement: Issues and Practice*, 20(4):19-27. https://doi.org/10.1111/j.1745-3992.2001.tb00072.x

Cohen, J.W. 1988. *Statistical power analysis for the behavioural sciences* (2nd edition). Hillsdale, NJ; Lawrence Erlbaum Associates.

Cordingley, P.; Bell, M.; Rundell, B.; Evans, D. & Curtis, A. 2003. The impact of collaborative CPD on classroom teaching and learning: An EPPI systematic review. In: *Research Evidence in Education Library*, Version 1.1. London: EPPI-Centre, Social Science Research Unit, Institute of Education.

Davis, J. & Martin, D.B. 2008. Racism, assessment, and institutional practices: Implications for mathematics teachers of African-American students. *Journal of Urban Mathematics Education*, 1(1):10-34.

DBE (Department of Basic Education). 2011. *Curriculum and assessment policy statement (CAPS): Senior phase mathematics, grades 7-9.* Pretoria: DBE

DBE (Department of Basic Education). 2013. National curriculum statement: Curriculum and assessment policy statement. Pretoria: DBE.

DBE (Department of Basic Education). 2016. *National senior certificate examination, 2015: Diagnostic report.* Pretoria: DBE

Dunst, C.J.; Bruder, M.B. & Hamby, D.W. 2015. Metasynthesis of in-service professional development research: Features associated with positive educator and student outcomes. *Educational Research and Reviews,* 10(12):1731-1744. https://doi.org/10.5897/ERR2015.2306

Gregory, K. & Clarke, M. 2003. High-stakes assessment in England and Singapore. *Theory into Practice,* 42(1):66-74. https://doi.org/10.1207/s15430421tip4201_9

Haberman, M. 1991. The pedagogy of poverty versus good teaching. *Phi Delta Kappan,* 73:290-294.

Heyneman, S.P. & Ransom, A.W. 1990. Using examinations and testing to improve educational quality. *Educational Policy,* 4(3):177-192. https://doi.org/10.1177/0895904890004003001

Julie, C. 2013. Can Examination-Driven Teaching contribute towards meaningful teaching? In: D. Mogari, A. Mji & U.I. Ogbonnaya (eds.), *Proceedings of the 2013 ISTE international conference on mathematics, science and technology education: Towards effective teaching and meaningful learning in mathematics, science and technology'* Pretoria: UNISA Press. pp.1-14.

Julie, C.; Mbekwa, M. & Simons. 2016. The effect size of an intervention focusing on the use of previous national senior certificate mathematics examination papers, *Journal of Educational Studies,* 15(1):1-20.

Ladson-Billings, G. 2000. Fighting for our lives: Preparing teachers to teach African American students. *Journal of Teacher Education,* 51(3):206-214. https://doi.org/10.1177/0022487100051003008

Ladson-Billing, G. & Tate, W.F. 1995. Towards a critical race theory of education, *Teachers College Record,* 9(1):47-68.

Maaß, K.; Reitz-Koncebovski, K. & Billy, G. (eds.). 2013. *Inquiry-based learning in mathematics and science classes.* Freiburg, Germany: University of Education Freiburg.

Martin, D. 2009. Researching race in mathematics education. *Teachers College Record,* 111(2):295-338.

NRF (National Research Foundation). 2011. *Concept Paper: FirstRand Foundation South African Maths Education Chairs Initiative.* Pretoria: National Research Foundation.

Pong, W.Y. & Chow, J.C. 2002. On the pedagogy of examinations in Hong Kong. *Teaching and Teacher Education,* 18(2):139-149. https://doi.org/10.1016/S0742-051X(01)00059-2

Popham, W.J. 1987. The Merits of Measurement-Driven Instruction. *Phi Delta Kappan,* 68(9):679-682.

Ruthven, K. 1994. Better judgement: rethinking assessment in mathematics education. *Educational Studies in Mathematics,* 27(4):433-450. https://doi.org/10.1007/BF01273382

Shepard, L.A. 2000. The role of assessment in a learning culture. *Educational Researcher*, 29(7):4-14. https://doi.org/10.3102/0013189X029007004

Shepard, L.A. & Dougherty, C.K. 1991. Effects of high-stakes testing on instruction. Paper presented at the annual meeting of the American Educational Research Association, Chicago, 3-7 April 1991.

Suurtamm, C.; Denisse, R.; Thompson, D.R.; Kim, R.Y.; Leonora Diaz Moreno, L.D., Sayac, N. & Vos, P. 2016. *Assessment in mathematics education: Large-scale assessment and classroom assessment.* ICME-13 Topical Surveys. https://doi.org/10.1007/978-3-319-32394-7

Swan, M. & Burkhardt, H. 2012. A designer speaks: Designing assessment of performance in mathematics. Educational Designer. *Journal of the International Society for Design and Development in Education,* 2(5):1-41.

Tate, W.F. 1993. Advocacy versus economics: A critical race analysis of the proposed national mathematics assessment in mathematics. Thresholds in Education, 19(1-2):16-22.

Teese, R. 2000. *Academic success and social power: examinations and inequality.* Melbourne: Melbourne University Press.

Van Geel, M.; Keuning, T.; Visscher, A. & Fox, J. 2016. Assessing the effects of a school-wide data-based decision-making intervention on student achievement growth in primary schools. *American Educational Research Journal*, 53(2):360-394. https://doi.org/10.3102/0002831216637346

Wall, D. 2000. The impact of high-stakes testing on teaching and learning: Can this be predicted or controlled? *System*, 28(4):499-509. https://doi.org/10.1016/S0346-251X(00)00035-X

Walter, C. & Briggs, J. 2012. *What professional development makes the most difference to teachers?* Oxford: University of Oxford; Department of Education.

Watson, A. & Mason, J. 1998. *Questions and prompts for mathematical thinking.* Derby, UK: Association of Teachers of Mathematics (ATM).

Wayman, J.C. 2005. Involving teachers in data-driven decision making: Using computer data systems to support teacher inquiry and reflection. *Journal of education for students placed at risk*, 10(3):295-308. https://doi.org/10.1207/s15327671espr1003_5

Wideen, M.F.; O'Shea, T.; Pye, I. & Ivany, G. 1997. High-stakes testing and the teaching of science. *Canadian Journal of Education*, 22(4):428-444. https://doi.org/10.2307/1585793

Yu, L. & Suen, H.K. 2005. Historical and contemporary exam-driven education fever in China. *KEDI Journal of Educational Policy,* 2(1):17-33.

www.ingramcontent.com/pod-product-compliance
Lightning Source LLC
Chambersburg PA
CBHW062031210326
41519CB00061B/7434